Interactive
Message
Services

Interactive Message Services

Planning, Designing, and Implementing Videotex

Dimitris N. Chorafas

McGraw-Hill Book Company

New York   St. Louis   San Francisco   Auckland   Bogotá
Hamburg   Johannesburg   London   Madrid   Mexico City
Montreal   New Delhi   Panama   Paris   São Paulo
Singapore   Sydney   Tokyo   Toronto

Library of Congress Cataloging in Publication Data

Chorafas, Dimitris N.
    Interactive message services.

    Includes index.
    1. Videotex (Data transmission system) I. Title.
TK5105.C5 1984   384   83-25580
ISBN 0-07-010850-1

1234567890 DOC/DOC 8987654

ISBN 0-07-010850-1

*The editors for this book were Stephen G. Guty and
Galen H. Fleck; the designer was Naomi Auerbach; and
the production supervisor was Reiko F. Okamura. It was
set in Baskerville by Byrd Data Imaging Group.*

Printed and bound by R. R. Donnelley & Sons, Inc.

Contents

Chapter 17. The Pronto System Software 280

Major Improvements   280
Toward an Electronic Teller   282
Touch-Sensitive Screens   286

Chapter 18. The Videotex Experience of German Banks 289

Preface

By the early 1990s we will make no distinction between voice communication and *videotex*, because all the facilities we will need for our work and pleasure—visual as well as oral—will be integrated into one system. The time is not yet ripe mainly because we still conceive of telephone conversation, text handling, and data processing as distinct entities.

Yet major changes are coming. An estimate by the National Science Foundation is that, by the end of the century, 40 percent of American households will be using two-way videotex services. That will lead to changes in economic and social organization as great as those made by mass ownership of motor cars and televisions. Home videotex machines will allow users both to reach into enormous databanks stored in computers and to communicate their own information, needs, and opinions directly to manufacturers, providers of services, and politicians. These videotex machines may be the most important products of the new *information technology*, which is now being developed by combining communications and computers.

Other predictions are that both videotex and *teletext* will be operating profitably in U.S. cities by 1990. The pace is picking up. In 1982, eight companies were testing teletext systems, whereas a year earlier only a few technical experiments were under way. By the same count, 14 videotex systems were being tested, mostly by banks, or were already operating on a permanent basis. That is almost double the number in 1981. Such experiments are bringing closer the day when people will use television screens in their homes to shop, bank, read about traffic conditions, get the latest stock market figures, and access other information. In the future, videotex will act like a magnet drawing many entrepreneurs with diverse information to attract consumers.

This book deals with a subject that is in transition; its goal is to bring together the *interactive message services* presently at our disposal and their possibilities of use. That is a beginning of integration. Here we are not talking of tomorrow's technologies; we are talking of what is available and running today. What we are missing are the images of the far-out applications, and they are what the present book aims to offer. It is written for middle management, the people who must work with the interactive message service. That is why it keeps away from theory and instead stresses practical examples. The power of technology is to create systems which can be put to profitable use. The book shows how and why.

Chapter 1 is introductory. Its aim is missionary work about the new facilities supported by computers and communications and what they can offer. Chapter 2 deals with the fundamental notion of the new executive services. Chapter 3 stresses managerial productivity and shows how the cost of the executive workstation can be justified. Chapter 4 discusses electronic mail and videotex, the latter being a more sophisticated version of the former. How can they be used to the organization's advantage? What may they be worth?

Standardization is the subject of Chapter 5, with particular emphasis on the North American PLPs and the CEPT standard. Chapter 6 reveals little-known facts about implementing a videotex system: timetables and the budget. The next two chapters, 7 and 8, aim to help the reader project an interactive system fitting his requirements. They explain the role of videotex, guide his hand in designing the applications environment, and give advice on planning premises. Such advice comes straight from practical experience.*

Chapter 9 is addressed to the evolution from mainframes to *viewdatabases*. It deals with a key subject which can make or break the interactive communication effort. Interactivity requires a secure and efficient access to the viewdatabase, but how can it be assured?

The aspect of internal control that troubles executives is the

---

* I know from long experience that an increasing number of women hold important positions in the computer and communications fields and in all kinds of enterprises: they are engineers, managers, and entrepreneurs—to say nothing of users, consumers, and customers. But English lacks a suitable neutral pronoun, and "his or her" is simply too stilted and cumbersome to use more than, say, once. So please accept such expressions as "filling his requirements" as applying to either male or female readers.

increasing dependence on computers for operational effectiveness as well as for financial reporting. Man-information communications often get structured as a result of chance and through many mutations. Now this structure must be planned.

Information is seen by most executives as an integral part of the management process. It is a key management responsibility to edit this information properly, as Chapter 10 emphasizes. The editing must be done in a consistent and able manner with the end user in mind (Chapter 11), and it should involve security and protection mechanisms in order to decrease the internal control risks.

The next two chapters, 12 and 13, focus on the transition from mainframe databases to the viewdatabase. Although editing terminals help introduce text, data, and graphics to the viewdatabase, manual introduction of all the infopages that management requires is unthinkable. The storing function must be carried out automatically, and the technology to do the job—and do it well—is available.

The following five chapters are dedicated to learning from experience. Chapter 14 brings the reader's attention to the software needs for the widening applications environment. Whether in banking or in industry, videotex frees the user from the weight of paper records; gives him *real enough time* actuality; makes it easy to access information; does away with specialist support; acts as a cost cutter in data processing; and provides all that in a user-friendly manner—if we know how to exploit the opportunity.

Chapter 15 is devoted to three case studies. The first presents the custom-made developments in videotex by the London Stock Exchange. The second has to do with the procedural work which a leading construction company found necessary to set up. The third is a scenario written to exploit videotex applications through an in-house system installed in a bank.

Home and home-office banking systems are the theme of Chapter 16: Pronto for retail banking and Chemlink and Banklink for wholesale banking. The Pronto system is further evaluated as to its software understructure in Chapter 17. Included is a discussion of a device which will most likely interface with the end user by the middle 1980s: the touch-sensitive screen. Finally, Chapter 18 considers the videotex experience of the German banks, the range of services offered to the end user, and a lesson learned from standards implementation in accord with the CEPT agreement: the cost of standards.

Whether it is called Viewdata, Datavision, Pronto, or Bildschirmtext, videotex is a product of technology and brings into perspective the interactive capabilities we will have available before too long. We are living in a competitive environment. Soon there will be many offerings in this field. Which one should we choose? Which way should we go? We know that we must upgrade management information. The tools for doing a neat job are ready and waiting. Some companies are already doing it; others must still learn how. This book is written to help them.

Let me close by expressing my thanks to everybody who contributed to making this book successful. From my colleagues, for their advice; to the organizations I visited in my research, for their insight; and to Eva-Maria Binder for the drawings, the typing of the manuscript, and the index.

Valmer and Vitznau   DIMITRIS N. CHORAFAS

Interactive
Message
Services

Survival in a Changing World

The more science expands the universe, the more it shrinks man. EISELEY'S POSTULATE

When wisely used to promote further development, science and technology can be beneficial. Not only do their long-term benefits more than balance their short-term reversals, but companies threatened by strong competition cannot let themselves become fat, complacent, and unresponsive. They must develop new processes and new products if they are to grasp the business opportunities of the future.

Any scientific discovery or invention, such as electricity, photography, telephony, the transistor principle, or the computer, is only a base upon which a new technology must be built before the field can be opened commercially. A product or process has little or no business value unless its manufacturing, marketing, and end use is economically acceptable: A demand must exist or be created before a product can be exploited in a profitable way in the marketplace.

At any given time, advanced technology has the characteristic that many years pass before the manufacturing stage is reached and marketing results can be evaluated as to profit and loss. In the instance of photography, some 112 years went by before the camera was produced commercially; its invention in 1727 reached a business conclusion only in 1839. But this transition period has shortened amazingly. In the case of the transistor the time lapse was only 5 years. The transistor was invented in 1947; its uses are now a legend. Among other things, it made possible big computers, microprocessors, satellite communications, the moon program, and a galaxy of new consumer products.

The transistor's main advantages, small size and cheapness, were improved dramatically by the invention in 1958 of integrated circuits, which now enable over 100,000 transistors to be placed on one tiny silicon chip. In 1969 the invention of the microprocessor made it feasible to put all the functions of a computer on just one chip, which led to the computer on a chip.

Office automation rests on two pillars: bringing the power of a microcomputer to individual desks at an affordable cost and applying new, imaginative solutions to meet old needs. The multifunction workstation, along with its microfiles, becomes cheap enough to be dedicated to a single user; electronic mail becomes a reality; interactive solutions such as videotex spread over the country; common carriers offer voice store and forward; and before long we will have voice input to and output from the computer system.

Other examples of the positive effects of science abound in all walks of life; a special case is biology. The structure of genes was discovered in 1953. It proved to be very simple, and that allowed scientists to probe the secrets of life from their smallest details upward. On a practical level, this led in 1969 to the inception of genetic engineering, which is already spawning a new industry just as the transistor did.

Science and technology expand human horizons, but they are also demanding partners. We must increase our effort on our own behalf: Lifelong learning, ingenuity, imagination, hard work, and the wish to keep on moving are the ingredients of this new way of looking at life. If technology knows no frontiers, we must behave accordingly. If we do not, we run the risk of verifying Eiseley's postulate.

New Values

It is an age-old principle that knowledge is power. A society can now operate only by having information open to everybody and getting each individual to respond via some market system. Information and, most particularly, knowledge have these special characteristics which contrast them to physical assets:

- The more we share them, the more of them we have.

- If we stop sharing them, they decay and eventually become negative assets.

Yet the sad thing is that no age has ever devised as many disincentives to personal effort, innovation, and productivity in such a short period of time as our own. We have reached the end of a long line, that of the industrial revolution.

In about 1830 the industrial revolution started in England, but food production still dominated work force allocation. The next 50 years saw exciting developments; the railroad, telephony, and electric light changed the way society looked at itself. The smokestack industries and the labor movement were in their beginnings.

1910 marked a real turning point in economic and social development. Rural population began to decline. Machine-based mass production was begun (1909, Model-T Ford). Telephony reached reasonable maturity. Shorter work hours were instituted, and society searched for a new base on which to rest.

1955 was the next turning point. Computers were commercialized (1953, first Univac installed at General Electric). White-collar workers exceeded blue-collar workers in number in the United States, and technology began to pick up speed. It culminated, in 1969, with men landing on the moon.

1970 was the following milestone. The microprocessor had been invented (1969, Intel and TI). The minicomputer became attractive to business and industry and paved the way for the distributed information systems concept. Mass production of computer gear led to fast cost reduction.

The end of the 1970s decade has seen the domestication of the computer (personal computers, videotex), the merging of voice, text, data, and image, the reduction of labor through robotics, the disappearing of classical work posts, and labor displacement on a worldwide scale (the Japanese invasion). In 10 short years Japan, without sources of energy but with a clear plan and orderly implementation, has become a worldwide industrial power and now moves toward the postindustrial age.

This is significant, and its impact goes beyond the temporary or even permanent loss of some section of the world's industrial market. Japan through postindustrialization and England through deindustrialization underline the substitution of current structure by the *knowledge revolution*. The new environment imposes tough prerequisites which we must meet. Nothing less than an imaginative combination of the following can solve our present and prospective social, cultural, economic, and industrial problems.

1. The informatization of our society

2. A 4-day workweek (or 3-week workmonth) to make feasible lifelong part-time education for all—from cradle to tomb

3. Leadership in the developing technologies

4. Emphasis on the research and development effort

5. The abolition of compulsory retirement of anybody not yet senile to preserve the developed store of knowledge

6. An orderly transition to the leisure society

Advancing technology increasingly implies multiple changes in our way of thinking. Education offers an example. The 1980s will see the passage from collective to individual instruction through interactive computer devices, the conversion of passive instruction into active instruction because of the availability of microprocessor-supported media, the implementation of far-reaching educational schemes on a distributed basis, and the substitution of school years blocked together in early life by lifelong learning.

People-oriented drawbacks can be overcome only by imaginative approaches and people-oriented applications. At the same time, the strength of a nation hinges on an ability to shape sound strategies which target economic winners. We have always thought of development as strictly economic. Now we have come to recognize our error: Above all, development is human and sociological.

Developments in Information Technology

By 1984 information systems had entered a new epoch following years of rapid technological change: microcomputers, data communications, databases, the implementation of multimedia networks that handle data, text, voice and image, and the wide dispersal of multifunction terminals that interface with data and text file processors. The new orientation stresses:

- Text and databases

- Online interactive services (including journaling and security)

- New approaches to voice communications

The trend from batch to online applications has accelerated, and a second drive is underway. We are moving from classical realtime toward interactive solutions with graphics and color supported by intelligent workstations. Yet a total conversion will take years to implement because of large software investments in batch programs, the multitude of prerequisites implied by online solutions, and the lack of system specialists able to supply those prerequisites.

As an integral part of the coming evolution, networks will support text, data, and graphics solutions, voice store and forward, and voice input/output (to and from computer devices) within the broad concept of office automation. Evolution in computers and communications will see to it that each decade is characterized by areas of application which reflect both the tools that technology makes available to us and the images that we have developed in exploiting them:

- From the 1950s, computing and processing were the basic frame of reference.

- In the 1960s, operating systems (OS) and memory storage attracted most of our attention.

- The 1970s brought into perspective distributed information systems (DIS) with distributed data processing, databasing, and data communication (DDP, DDB, DDC).

- The keyword of the 1980s is the intelligent workstation (WS) based on personal computers (PC) communicating through local area networks (LAN) that link distributed databases together.

- Most likely, in the 1990s artificial intelligence (AI) will be the focal point. It will lead to expert systems that have learning capability and are able to respond to environmental stimuli.

These developments have had great impact on management functions. They have made feasible novel ways and means of improving both mental and clerical productivity, of bettering a firm's competitive position in the marketplace, and of keeping costs under control. But they require an appreciation of what technology can offer, and they call for skill and knowhow in their implementation.

Rule 1 with modern computers and communications systems can be expressed as follows: *The more advanced the technology we use, the greater the degree of necessary foresight, insight, and preparation.* Those three basic ingredients of advanced systems still call for the highest level of skill. The more we use advanced technology and hope to gain competitive advantages from it, the more we must focus on the preparatory work and the better brains we must use in executing it. The tools are here. What about the *knowhow?*

The workstation, with its self-contained computing power, is cheap enough to be dedicated to a single user. The software for handling text, number processing, and graphics is available to a substantial degree through packages. But the preparatory work still needs to be done. To satisfy that requirement, we must train both the users of the new systems and the specialists projecting them. Without that dual action we are simply spinning our wheels.

Figure 1-1 makes this point by putting in perspective the integration, through architectural knowhow, of physical and logical components into a working aggregate. This aggregate is a distributed information system supporting databases at various levels and, most importantly, focusing on user interfaces.

The systems which we presently build have only one purpose: to aid the end user. To do so, they absolutely need the end user's participation. In turn, this participation is promoted both by the existence of new images in terms of services and by top management's decision to move ahead.

What is necessary is a vision of the direction in which we should be moving, and it can best be reached through a steady research effort not only inward in terms of system design but also outward by observing what takes place in our environment. The major findings on high technology from a research project which I conducted among leading American, European, and Japanese industrial firms and financial institutions is summarized in the following paragraphs.

First was confirmation of the trend toward the application of personal computers, local area networks, and videotex services in financial and industrial environments. There was, for instance, confirmation of strategic perspectives in the banking industry: home banking (HB) and electronic tellers going well beyond the automated teller machine (ATM) level. Home banking has reached its first level of maturity. Broadly, it divides into two basic solutions: (1) videotex supported through public and private networks and basically using the telephone lines and the TV set and (2) on/off intelligence provided through personal computers.

Second, there is an exploding PC market featuring more than 100 competing firms offering some 150 products. The top microprocessors becoming de facto standards are Motorola's 68000 series with 32 bits per word (BPW) and Intel's 80X86 family. The latter has a 16-BPW processor; it will be succeeded by the 80186, with twice the 8086 speed,

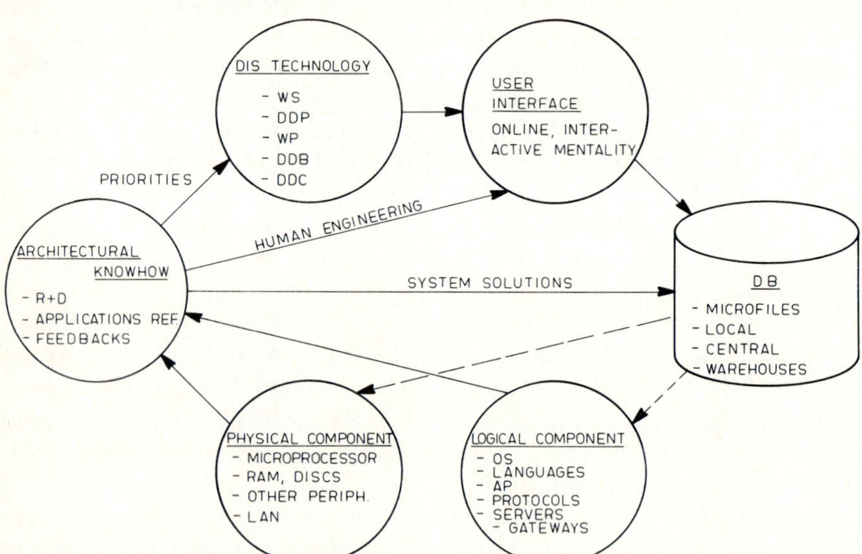

Figure 1-1 A way to integrate, through architectural know-how, the logical and physical components of an information system.

addressing 1 megabyte, (MB) and incorporating 55,000 transistors, and the 80286, with four times the 8086 speed, addressing 2 MB, and with 130,000 transistors on a chip. Other interesting Intel projects are the i432, the database (DB) the OS, and the transaction processors on a chip.

Third, there is steady evolution in and improvement of LAN architectures. The second generation came in early 1983. Many applications already in process are aimed at replacing the minicomputers. Top management is one of the areas in which LANs are used to advantage.

Fourth, since its beginning in March 1979 in England, videotex has taken hold and today has an impressive array of applications throughout the United States, Europe, and Japan. Among the best American applications are the following:

- *CBS/AT&T.*   It operates in Ridgewood, New Jersey, with two key viewdatabase (VDB) sponsors: Holt, Rinehart, and Winston; and W. B. Saunders.

- *Knight-Ridder/AT&T.*   Coral Gables, Florida. VDB by Viewdata Corporation of America and *The Miami Herald.*

- *Times Mirror.* Mission Viejo, Palos Verdes Estates, California. VDB by the Los Angeles Times Mirror firm, Bank of America, World book, and Recognition Activity.

These early experiments linked home video terminals to computerized databases; they provided information ranging from stock market reports and news to theater reviews and advertising. The systems can also be used for shopping and banking from home.

The CBS/AT&T offering has entered its second phase of testing. Much of the information content is provided by CBS Inc. AT&T furnishes the home terminals, communications lines, and computer facilities. Its new computerized frame creation terminal produces sophisticated graphics for videotex services. Artwork is contributed by the Bell Laboratories.

To promote home banking through videotex, the Bank of America has joined forces with the Los Angeles Times Mirror communications firm. The First Bank System of Minneapolis established its own experiment in the Dakotas and has been joined by local newspapers, department stores, book sellers, and other merchants. We will return to these experiments. Chemical Bank offers Pronto, which can be easily assimilated to a private videotex service for home banking.

Another quite interesting American videotex experiment was conducted by the Tandy Corporation and AgriData Resources, a Milwaukee publishing company. Its aim was to bring computers to the world of farming. Also directed at farming is the commodity information supplied by the AgriStar network. Through a public data carrier, farmers are able to determine the latest prices for pork bellies, corn futures, and soybeans on the Chicago exchange, and they are therefore able to plan their plantings and harvests more precisely. In addition, AgriStar offers the farmer the weather outlook for each and every county in the United States, as well as weather forecasts for foreign countries. The latter are helpful in predicting crop yields and the resulting commodity prices.

Not all videotex experiments have met with success. In England, despite enthusiastic predictions about the number of users it would attract, Prestel has stagnated at about 20,000 subscribers, many of them commercial or industrial, and it has cut services over the years. West German officials diagnosed Prestel's weakness as an inability to go beyond the system's own data banks to link users with computers of companies offering services through the system, such as stores, banks or mail-order houses.

The Japanese solution CAPTAIN (character and pattern telephone access information network) is promoted as a national service by Nippon Telegraph and Telephone (NTT). Similarly the German *Bildschirmtext* (Btx) was established by the German Post Office (Bundespost). It was due to operate nationwide in September 1983, but it was delayed at least 6 to 8 months. The delay angered some of the 1500 organizations—particularly the banks and mail-order houses — that expected to use the system to expand product sales. However, suppliers of peripheral equipment expressed relief because they were experiencing their own delays, and in some cases the managements of information providers looked at the delay as an opportunity to gain time in making up their minds.

In all probability there will be a nationwide installation of Btx in 1984, the original pilot test having been limited in the number of cities and participants. The Bundespost believes the infrastructure of the new information and communications system will be fully established after reaching only about 10 percent of all households.

Table 1-1 presents approximate user and information provider statistics for the videotex systems currently operating in Europe. Let us recall that most systems are still in the experimental stage; by and large they have given first-class results.

Most importantly, in February 1983 in Geneva, European and North American representatives signed a joint accord aimed at establishing a bridge between regional videotex standards. The North American presentation level protocol (NA/PLP), the Canadian standard, has

Table 1-1   Users of and Information Providers to
the European Videotex Systems

Country	System	Number of users	Number of IPs
Finland	Telset	600	30
France	Teletel	2,500	200
Germany	Bildschirmtext	5,200	2,250
Great Britain	Prestel	20,000	170
Holland	Viditel	11,000	140
Italy	Videotel	1,000	45
Spain	Videotext	400	100
Sweden	Datavision	1,000	40
Switzerland	Videotext	2,000	100

been linked to CEPT, the new European standard, through a "super-group" switching organization.*

Videotex, personal computers, and local area networks see to it that the end user can reach, directly online, a wealth of information supported through a network of databases and consisting of:

- *Microfiles* (personal files)

- *Company proprietary data* (text and data *are* a company resource)

- Other information available through publicly offered *text and data warehouses* accessible at a fee

Slowly but surely, mainframes will be phased out of processing. Increasingly, processing will be distributed (LAN/PC) and the mainframe will remain as a text and data warehouse and central switch. The data communication nodes and microfiles also will be distributed, with personal computers increasingly able to run hard disks.

Private LANs and Public Services

We have been speaking of workstations, videotex, personal computers, local area networks (LAN), graphics presentation, and the impact that high technology can have on financial institutions and industrial firms.

---

*In reality, CEPT is a misnomer for a videotex standard. The European Committee for Post and Telecommunications (CEPT) is based in Geneva, Switzerland, and acts as an advisory and coordinating body. One of the recommendations it has made is the so-called CEPT standard for videotex.

One way to divide these services is between private and public. The WS, PC, and LAN will tend to fall in the former class; videotex in the latter. The dichotomy is not, however, absolute.

We can run a videotex service on a LAN. Cluster/One, the Nestar/Zynar offering, is doing it through a package built by an independent software firm. We can run electronic mail on LANs and PCs; Omninet and Cluster/One are examples. But videotex and electronic mail are also public offerings par excellence. Most value-added networks (VAN) support electronic mail, and in the long run videotex is meaningful only if run through long-haul networks also.

What personal computers, local area networks, and videotex have *in common* is that they are the foundations of:

1. *The offering of highly competitive low-cost services to a firm's clients*

2. *The automation of office work at the company itself in order to reduce costs and increase our ability to survive*

But the implementation of new technology requires new images and new concepts. It cannot be done through the electronic data processing (EDP) approach we have known since the 1950s. If management decides to put the organization in the path of slow death, the "time-honored" EDP approach is the way to do it. If management opts for survival, then quite the opposite way should be chosen.

Figure 1-2 dramatizes this observation. Has management decided to be early in technology and reap the benefits derived from it? Then, it should be focusing on point *A* of the curve. Or has management decided to stay behind and be a late follower? In that case, point *Z* is fine. But if management does decide to drive for point *A, then* the concepts, the tools, and the way to implement them have to change. Success against competition cannot be had with outmoded ideas and tools.

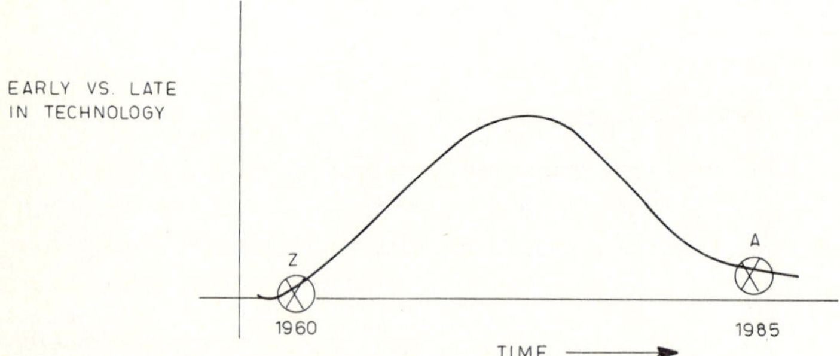

Figure 1-2 Banking and technology: the basic choice. Early vs. late implementation of information technology can have a dramatic impact on the competitive position of an organization.

Not long ago, a certain financial institution decided to implement personal workstations based on PCs and LANs. But after the first phase of the project had been completed and some of the programs were ready for implementation, it was found that the PCs could not be integrated into the workplaces of the employees. This came by way of a "discovery" that there were, at the foreign operations sector of the branch office, 25 typewriters and a large number of calculators used by the sector's 30 employees in producing different documents.

Clearly it is a nineteenth century approach to use typewriters. But the example also suggests that the systems study did not tackle the fundamental problem. If we wish to improve productivity, we have to radically change the approach. The fact that the plurality of electromechanical equipment does not allow the integration of PC and LAN at the workstation of the foreign operations employee—*which should be rule 2*—leads to the necessity for the PC to sit like an orphan next to the aged nonintelligent computer terminal that had served the office for 10 years.

The "orphan principle" is exactly what is needed if we wish the application to fail. This way of thinking is obsolete. Three reasons:

- The microcomputer can help if it becomes a personal workstation.

- By depriving the employee of the microcomputer support, we essentially kill the PC's reason for being—and with it the project.

- By putting the PC and LAN on equal footing with the old terminal, we strangle them. Figure 1-3, a well-known finding of applied experimental psychology, shows why.

With all new media and new procedures there is the learning curve of Fig. 1-3. The key factor here is time. The nonintelligent online computer terminal is something the employees already know. PCs and LANs are new to them; in the first phase, they were programmed to cover only one-third of the old terminal operations.

The point never to be overlooked is that it is useless to spend money on new technology if we hold onto the concepts of the old technology. Let me amend that statement. *Such a policy will be not only useless but also destructive.*

Let us take a constructive approach. On the supposition that management really decides to move the organization forward and keep it ahead of competition, Figs. 1-4 and 1-5 present the old approach and the systems solution to complex operations at the branch office. (The example is taken from the banking industry to present the positive side of the example given in the preceding paragraphs.)

In the old approach we had calculators, typewriters, marking equipment, stand alone data processing (DP) devices and DP terminals. Over and above that we had duplicate data entry activity at the center.

The new systems solution should be to integrate data and word processing and, therefore, the hitherto separate functions of word processor, calculator, DP terminal, and data entry facility together with such new activities as information enrichment, error control, calendar services, and graphics. All that must be done online and interactively by the professional employee. The information is captured only once and used many times. That's *how* and *why* we can improve client service *and* reduce clerical costs.

Such a systems solution has to be either fully adopted in order to bring benefits or not implemented at all. Half-baked approaches are no different, so far as results go, than the man who has his feet in an oven and his head in a refrigerator and thinks that on the average he should be comfortable.

Managerial Responsibility

The emphasis in the preceding example is on *preparation* and *planning*. This has to be done at all levels—governmental, corporate, personal— and it must be based on a well-thought out, valid methodology. "The plan," President Eisenhower once said, "is nothing. Planning is every-

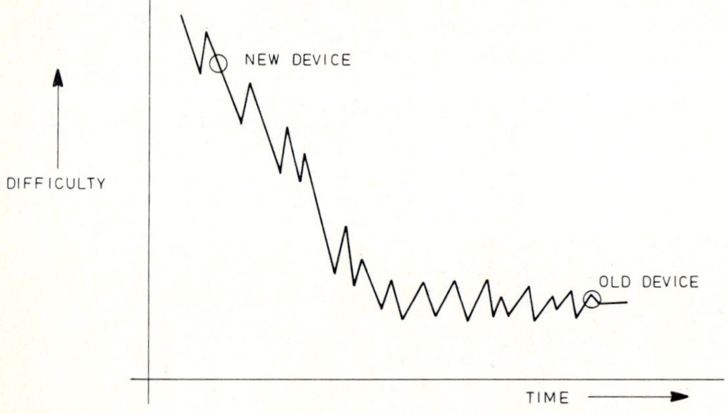

Figure 1-3 Depending on the level of user difficulty with new systems, equipment, and processing, the introductory phase of any information system will follow this learning curve. The curve has been found, by applied experimental psychology, to be characteristic.

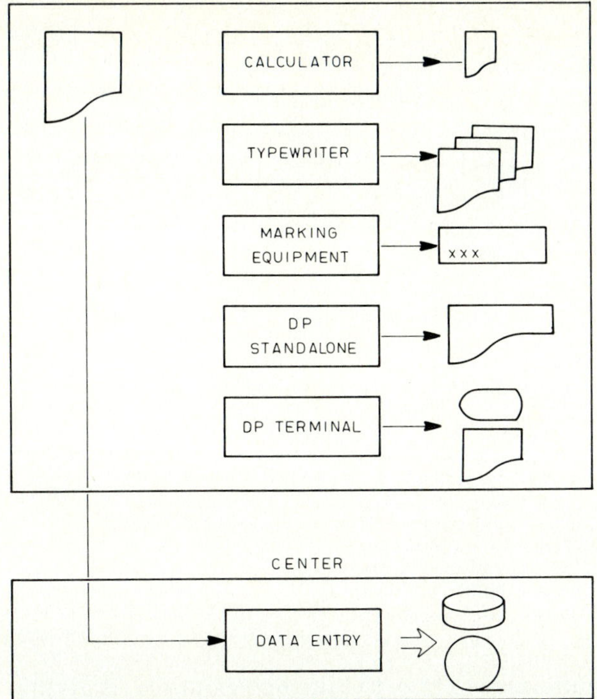

Figure 1-4 The old approach to complex operations at the branch office (1950-1980) was to introduce the computer and peripherals and leave discrete islands of manual- and electromechanical-equipment-oriented operations.

thing." The cutting edge of technology substantiates that observation. Governments should surely know that the high degree of advanced science which is involved—and the many fronts to be faced at the same time—call for huge amounts of money to be raised and properly allocated and the capture of at least 10 percent of the world market for a given product in order to feed a polyvalent industry working on the fringes of technological advance.

To establish a reliable base in a marketplace more rewarding and more competitive than ever, we must build a knowledge-intensive strategic plan. We will not do it so long as we remain undecided about whether to be guided by trade-union-financed parties protecting crafts that became uncompetitive 20 years ago or by parties financed by the great manufacturing companies whose businesses should be transplanted to third world countries tomorrow.

What are we doing about tomorrow? How does the trend look? The

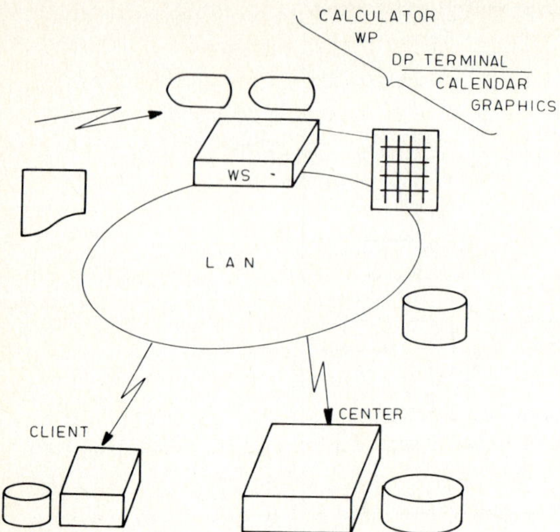

Figure 1-5 The new systems approach to complex operations at the branch office (1980 to end of century) integrates at the workstation all the facilities vital to mental and/or clerical productivity.

European Economic Community (EEC), to take one example, is giving the British farmers $240 million a year to make up for export losses in the farmers' lamb war with France. At the same time, EEC has allocated less than $40 million in its computer industry budget and still has to come up with a coherent information technology plan. In Europe as in America, billions of dollars are poured into the ailing industrial sectors of yesterday. As a result, no significant government support is being given to telematics, the industries of the future.

Japan has put great emphasis on *robotics,* and it alone accounts for more than half of the installed equipment. *Telematics* can be our chance.* An office without employees, a factory without workers does not mean that either is totally without people. As office automation and robotics assume the tasks which were formerly assigned to white- and blue-collar labor, workers and employees must be transformed to better-trained elements which have higher functions. We should not only overcome the problem of human obsolescence but also carry men and women to a level of intelligence we have never known before.

---

*"Telematics" is the word commonly used in Europe for the integration of voice communications with text, image, and data communications. In the United States, "compunications" is often used.

The new technology is a two-edged sword: It relieves people of manual labor, and it also obligates them to keep themselves in a state of steady evolution. Only by updating their skills can they avoid becoming the marginal or even unwanted elements of a fast-developing world. If people are not prepared, the consequences will be grave: with a badly utilized technology, competitiveness is lost, and costs do not have a corresponding benefit.

For good reason the National Science Foundation conducted a research aimed at establishing the factors responsible, during the preceding 50 years, for United States economic progress. According to its report, 50 percent of the growth of the United States economy is a direct fallout of R&D and innovation. Industries which have concentrated their effort on technology have twice as high productivity, three times faster business growth, and only one-sixth the increase in costs of traditional industries over the same time period.

The industrial prospect for rich countries is that the old big-company, trade-unionized, smokestack industries—steel plants, car plants, and the like—should be moving out of the mainstream of their manufacturing and into knowledge-intensive tasks. We must appreciate that the following are characteristic of all developed countries:

1. A sharp turn to skilled jobs requiring *higher capital investment* in new and growth lines of business

2. A fast-growing need to *manage the productivity* of key resources: knowledge, time, and capital in that order

3. An evolving fact of life: that capital investment has become *inseparable* from the need for more *knowledge work*

4. A not yet conscious realization that the productivity of people requires *lifelong, continuous learning*

5. A basic, but not appreciated, reality that *knowledge* has to be *applied to all resources:* time, capital, key physical elements, and knowledge itself

6. The overriding need for a *properly projected* and well-coordinated *far-out plan*

We cannot match the longer-viewed Japanese with piecemeal, uncoordinated programs. In moving ahead toward new, largely unknown fields of endeavor, such as telematics, there will be accidents, but accidents help the prepared mind. Like all innovators, we have to be risk takers; we must be willing to make mistakes, recognize them, and learn from them. We cannot stand still, behind ill-fated protectionist barriers, while others erode our markets and take away our opportunities. If we do follow such an unwise policy, sooner rather than later we will pay for our shortsightedness.

Growth Industries

Here is some eloquent data concerning the growth industries: The United States currently has about 50 percent of the world's installed computers and communications systems, compared with Japan's 12 percent. At this time, the Japanese computer companies control three-fourths of their home market—giving them 7.5 percent of the world market—but there is a long-standing plan by which Japan Inc. hopes to reach about 20 percent of the world market by 1986.

This strategy focuses on the growth industries and dates in its conception back to the late 1960s. The first 5-year plan covered 1970 to 1974; it has as an objective the plug-compatible mainframes. Financed to the tune of 50 percent of all R&D expenditures by the government, the effort primarily involved Fujitsu and Hitachi (then an RCA licensee). The result was the M Series launched in the United States and then the world market through the Itel (for Hitachi) and Amdahl (for Fujitsu) Trojan horses.

IBM rightly guessed, in the early 1970s, that a major danger to its installed base would come from the Japanese. But until recently the instrumental party was not known; Itel's Advanced System (AS) was thought to be a largely American breakthrough. In the meantime, Japan's second 5-year plan was not only initiated but also completed. It covered the 1975 to 1979 time frame, and its fruits were in semiconductor technology.

The timing has been perfect. The semiconductor crunch of 1979 gave the Japanese manufacturers their chance to carve out a hefty 42 percent of the United States market in 16K chips (K = 1024; see Glossary). And while such major advances are being made in the United States, and to a greater degree in Europe, the information industry is still in search of a policy.

The Japanese third 5-year plan is going strong and for good reason. The same companies that participated in the semiconductor consortium—Fujitsu, Hitachi, Mitsubishi—are now pushing into software leadership. Again half the R&D budget is supplied by the government and the goal is not just to close the current gap, but to become front runners. Software is expected to replace hardware as the top income earner of the mid- to late 1980s.

This is the story of the *third-generation online systems* in which telematics will be pivotal. Let us make no mistake about it, telematics is to the 1980s what distributed processing was to the 1970s. Most people talk about it, few have it, and the leaders in its application will come away with the benefits. As we talked about it in this book, system integration should focus on ironing out the inherent difference between supporting the word processing task and supporting data processing at the intelligent workstation level. The systems effort will involve both some far-sighted solutions and some compromise.

When we refer to data processing, we mean applications programming, data entry, file retrieval, record editing, and file sorting. Word processing has its own prerequisites and these, too, must be in order to improve efficiency and increase output per worker hour. We need a system's approach to the workplace of tomorrow and a very good job of educating users in all that the hardware can do. Above all, as the Japanese experience demonstrates, we need precise goals.

Who will plan for an orderly transition to the new environment? An unplanned third industrial revolution could cause much misery—and endanger democracy itself.

- We have to strip our problems to their core in order to find valid solutions.

- We must prepare for the coming microprocessor-intensive environment and foresee the resulting social problems, rather than leave them to blind forces.

- We must make sense out of conflicting elements.

And nobody should try to exclude the microprocessor-based technology from products and manufacturing processes in order to preserve jobs. That would merely make products uncompetitive with those of the other advanced countries. Even if such an irrational solution were adopted, many millions of permanently unemployed would still remain because their acquired skills have been made obsolete by technology. Having failed for so long to update our skills, we are damning technology for "structural unemployment." The real difficulty is that our population is mentally aged and is insufficiently flexible to acquire new capabilities. Does anybody really want to provide new opportunities for the broader strata of our population? The well-being of society depends on the answer.

A technologically retarded country would have far greater balance of payments difficulty in keeping its unemployed at the standard of living to which they had become accustomed than would an efficient, fully automated economy. And that is where high technology can help.

2

Executive Services

*"The race is not always to the swift, nor the
battle to the strong, but that's the way to
bet."* DAMON RUNYON

Whenever we deal with technology in the office, we find that the real problems are behavioral, not technological. Perhaps no other aspect of the new computers and communications services impacts managerial acceptance as much as the effect on the traditional ways of doing business.

We do appreciate, for example, that voice and video teleconferencing services, in which telephone lines and communications satellite links are used, will increasingly replace person-to-person meetings and interstate travel. Memos and letters typed on paper will give way to computer-based electronic mail and calendar services displayed on videoscreens. Clumsy and bulky hardcopy filing will be replaced by videodisk-oriented systems for information storage and retrieval.

A variety of services will be integrated into compact, executive workstations, but wisdom based on the experience of the past three decades suggests that the office of the future will evolve slowly toward the end of this decade and beyond. The pace may accelerate if we are able to overcome the psychological barriers. Will tradition-bound managers who have already resisted some of the earliest, most rudimentary forms of automation be willing to work in the office of the future? To a very large extent, the answer lies in preparation and training. In general, it is the human factor that will determine the success or failure of office automation technology in the marketplace.

Much of the resistance to the office of the future concept is attributable to the "what's in it for me?" feeling. A significant part of both management and the clerical-secretarial workers worries about the impact of such unfamiliar new concepts as electronic mail, workstations, and personalized information processing. Managers are used to hardcopy files, typed messages, telephones, and all the standard office apparatus. And they are afraid the changeover will undermine their positions.

It has also been suggested that companies might have to wait out a generation or two of executives before fully achieving an office of the future. That is absurd. The generation gap in managers has existed since the beginning of the computer era some three decades ago. The reason for it is failure to communicate with and train and support the people directly involved in the changeover.

Management Functions

If we aim to support management in doing a more productive job, we should establish at the beginning the basic functions of the management profession. Although the details may vary from one position to another, the basic functions of management are six in number: forecasting, planning, organizing, staffing, directing, and controlling.

Each and every one of those functions is involved in decisions about the new information and communications services. Since the 1960s, computer and communications perspectives have been broadened not only by advances in technology but also, if not primarily, by the demands imposed on the managerial job. There have been unprecedented complexities of laws, market drives, product orientations, costs, profits, and labor sensitivities.

Computers have provided substructure for a fourth dimension in financial and industrial organization. Like the telephone, which enhanced the potential of management communications, the evolving integration of *text, data, image,* and *voice* promises to revolutionize the way business does things. Many of the classical line and staff functions will be integrated into the new information system through the instantaneous communication of the person who has the problem with the data needed to do the job.

To run an organization, a manager must have four capabilities: decision making, motor power, transmission, and control. From decision making to management control, enterprises require an extraordinary amount of conferencing and document handling, and that is exactly what we aim to improve with the now-evolving structure of computers and communications networks.

To meet such managerial requirements, a new industry is in the making, one that is potentially wider and more powerful than the computer and telephone business itself. IBM, AT&T, DEC, Xerox—and also Exxon, Wang, Datapoint, Olivetti, and Apple—are but a few examples of companies tooling up to operate in the different layers of this industry. Major organizations are preparing for services which are add-ons to the transmission facilities offered by the telephone companies or even direct, privately owned links.

But although communications is a vital part of management action, it is not the whole of it. Management plans are increasingly disrupted by:

- Accelerated time scales
- A compressed time in which to adjust the company's own forces and the client's views
- Organizational and structural developments, which go beyond the classical two-dimensional chart
- Toughening competition
- Accelerated product obsolescence
- Changes in the established standards
- The impact of a broadening geography (local, national, international)
- An unprecedented industrial (and product) breadth and range

By now we have ample evidence that those developments have a major impact on decision making and on the acute discernment needed as a basis for sound decisions. Computers and communications should help management forecast business evolution and thereby provide the executive with the lead time necessary from decision to application. But to take advantage of the capabilities provided by technology, we should develop management skill, and we must be able to comprehend *and* simplify complex situations.

As I will never tire of repeating, computers and communications can help management *only* if the right study is made from the beginning, enough lead time is allowed for the study, and management itself backs the study from the start. A system development cycle involves the definition of:

- User requirements
- User problem areas
- Project organization
- Logical design
- Message definition
- Databases

- Data communication

- Physical requirements

To support an internal control structure properly, the system must incorporate detailed document design, it must undergo a test plan, and it must be assured of user and operational documentation. The analysts should pay proper attention to the conversion plan, guarantee the existence of an audit trail, provide an uninterrupted flow of applications, detail how programs are to be operated, backed up, and recovered, identify allowable user commands and functions, and develop security and control education programs.

An interactive management environment is characterized by short input messages which, to a large extent, are queries. Three criteria are important: First, facility in implementing and user functions (simplicity, foregiveness, and low training requirement). Second, fast response time; delays are unacceptable. Third, agile database design and secure access.

When we talk of management applications (and most generally of WS usage) we always think of user-friendly interfaces. Yet we should also recall that user-friendliness is usually a response after the fact. The "fact" rests both on technology—query languages, menu approaches, and prompting solutions—and on the organizational work which should precede the implementation. We should not try to install a system without having a good grasp of our needs and of ways to meet our needs.

An interactive environment has many characteristics, but those which we are considering tend to outweigh others. When executive time is involved, no extensive queuing or processing time should be tolerated. Responses to simple inquiries should occur in less than 2s (seconds), and consistency of response time for similar queries is just as important as minimal delay.

We have spoken of security requirements. To serve management purposes, they must be much better thought out and more detailed as the system goes online. Two levels of entry control can be identified for a workstation:

1. *System entry.* It involves station, via modem, and sender/receiver, through badge and personal identification number (PIN).

2. *DB level of protection.* Trap number and closed user group at the design stage; author identification when in use.

Just as fundamental is the need for journaling and use statistics; the former for restart and recovery and also for auditing, the latter for system efficiency and dimensioning. In management information systems (MIS), auditing must be a steady and all-encompassing job. Texts and data are key resources, and the use of key resources should be steadily audited.

New Approaches to Text, Data, and Voice

Computers will complete the transition from the data processing center to the office only when they have been adapted to the users' specific needs. The keyboard is the classical tool in some occupations, notably secretarial work and reporting, but it is a nuisance in others. Alternatives are the graphics tablet, mouse, touch-sensitive screen, and voice data entry.

Graphic tablets are more flexible and user-friendly than keyboards, and their *templates* can be designed for specific workstations. The technology is ready. Graphic tablets (and/or touch-sensitive screens) can be implemented at the managerial as well as the day-to-day operational level for predefined and variable text and data and as graphics editors and instruments for color shapes and high-resolution graphics. They are low in cost ($250 to $750 per unit), and they can be attached to the personal computer at the discretion of the person at the workstation.

As for voice communications, usage patterns indicate that executives prefer voice messaging to the telephone. In terms of larger voice mail systems, AT&T and Rolm—among other vendors—have introduced voice messaging. Voice Store and Forward (VSF) and Phonemail are now offered by AT&T and Rolm, respectively, as options on their private branch exchanges (PBXs).

VSF is an "adjunct processor" to the Dimension PBX, to be used until the long-awaited all-digital Antelope system is introduced. Phonemail is also encased in a separate box in Rolm's computer-controlled business system. Both offerings have the same useful feature: A red indicator light on the handset lets the user know that a message is waiting. Messages can be received via Phonemail even if the handset is in use. In this sense, Phonemail is superior to a phone-answering machine. Delayed delivery is offered by VSF, but not by Phonemail.

Whether larger or smaller systems are being used, several communications possibilities are available to the users of voice mail. Through simple commands the systems now in development can:

- Accept messages from a variety of users
- Confirm the reception and identify the senders as well as the destination parties
- Store and forward the messages
- Repeat or suppress messages as instructed by the user

- Transmit messages instantaneously or through delayed response to one or more persons

- Add information such as date and exact hour

Having taken notice of the message, the receiver can instruct the system, through very simple commands on a touch-tone keyboard, to play the message, erase it, forward it to another party, add to it, and so on.

The computer conference is another example of the modern use of voice facilities. It is a continuous meeting in which a series of users are connected together. Each time a comment is made, it is automatically sent to all users. This permits a meeting to take place among users on an extended basis.

Computer mailbox and conferencing systems are suited to group interaction, which is now the most critical management problem in most organizations. Mailboxes are valuable for a wide range of tasks from coordinating meetings to receiving comments on a proposal or a report from people in different locations.

Still another development in voice communications is cellular radio service. A metropolitan area is divided into a honeycomb of cells each of which is equipped with a low-power transmitter. This permits the reuse of scarce radio channels. The caller, while moving across the city, is automatically passed from one cell to another.

At the same time, the use of satellite communications opens new horizons. Satellite use has so far been confined to telephone and television services that require dish-shaped antennas at least several feet in diameter. But solutions that have recently been proposed would allow individuals to communicate via satellites through small antennas.

The satellite would also be able to pinpoint the location of an individual or a vehicle, allowing help to be sent in an emergency. Mobile satellite, or Mobilsat, of Pennsylvania, has asked the Federal Communications Commission for permission to start a mobile telephone service using satellites. Calls from automobiles, or even from passengers in airplanes, would be transmitted to the satellite and back down to earth near the call's destination and tied into the phone system.

The National Aeronautics and Space Administration (NASA) is supporting efforts to establish systems such as Mobilsat. Canada is planning a similar system. As early as the mid-1970s NASA demonstrated the ability to communicate via satellite with devices no bigger than walkie-talkies and to track moving vehicles. In one test, eight ambulances were equipped with systems that enabled a patient's medical signs to be transmitted via satellite to the hospital and instructions for treatment to be transmitted from the hospital to the ambulance.

Such down-to-earth satellite implementations mesh nicely with cellular radio. One set of frequencies proposed for use by Mobilsat is now set aside for future expansion of cellular radio mobile telephone service. The effort is supported by numerous potential customers, including oil companies (who would use it to transmit data from remote rigs), search and rescue authorities, and rural telephone companies.

Polyvalent Voice Facilities

The use of voice can be polyvalent. Voice data entry, for example, lowers the costs of both data capture and training the user. It may also be the only choice for occupations in which freedom of movement is essential. But is voice data entry ready for practical application? For a structured information environment, the answer is yes! Lockheed has used voice data entry in a hybrid assembling area since 1979. It helps to track materials and stages of assembly and to enter product-assurance test data. A 48-word vocabulary includes digits 0 through 9, the letters of the alphabet, and such commands as "recall" and "kit."

The new generation of voice data entry systems will be user-friendly and aimed at managers and clerks, not just assembly line workers. Integrated circuits have helped in voice handling:

- From voice recognition and answerback to the improvement of the quality of sound

- In the development of intelligent telephone lines

One of the recent implementations (voice communication on the Renault 11) has a 128K RAM and a rate of 1 (kilobit per second), and hence a speaking time of about 2 min (minutes). Research aims to lower the rate to 0.8 kbps. Another project, by connecting 16 devices together, is to reach a vocabulary of 1600 words—more than the average citizen uses.

Voice recognition devices provide added security. TI and AT&T plan to use them to address the phone set for automatic call-up. TI and Renault project a "start motor" voice recognition. If anyone but the owner pronounces the words, an alarm is set off.

Portable analysis and synthesis speech (PASS) will revolutionize industry as learning machines get into the act. It will also lead to greater productivity and/or safety:

- A visual message is interpreted in 6 min by 75 percent of people.

- An auditory message is interpreted in 2 min by 99 percent of people.

These examples are based on current, not on far-out, technology, and they help explain to a degree why users look at data and word processing (DP/WP) and voice integration as fundamental needs. The concept is to have both voice and data operation over conventional telephone wire-pair facilities, coaxial cable local networks, radio links, and so on.

In the race to merge data processing with telecommunications, Northern Telecom's Display Phone packages a telephone, a modem, a computer, a display screen, and a pullout keypad. It sells for about $2000. The first model featured 32K memory, 11 pages of directory with nine names and phone numbers per page, automatic log-on for access timesharing services, a digital clock, a calendar, and a reminder service that sounds a tone to get attention. The unit has a full Qwerty keypad, battery backup that lasts for 3 days, a 103-type modem, a 300-bps (hits per second) fixed data rate, provision for setting half or full duplex, odd or even mark parity, and a 40- or 80-character display with 24 lines.

As the manufacturer commented, the $2000 price tag is no problem. The big question is compatibility. Will the Display Phone be able to support existing full-screen capabilities and programs made for applications which were originally DP-oriented? What about private branch exchange (PBX) compatibility?

Private branch exchanges are the gateways of the office to the public telephone system. The facilities they support include voice mail, which helps to dissociate the sender from the receiver and gives each a greater freedom of movement. Teleconferencing can be promoted through a PBX, and so also voice input and output can be. With the trend toward digital PBXs, quality transmission of voice will improve and the merging of data, text, image, and voice will be brought another step forward.

Automatic dialing, tracing, follow-me, line hold, line switching, use statistics, and accounting characterized the preceding generation of computer-based PBXs, but emphasis is now shifting to overall telecommunications optimization. A study by Chase Manhattan demonstrates that investments made in intelligent telephone gear today are the best way to keep the ever-growing telecommunications costs down tomorrow.

Ultimately, the most expensive part of the telephone system is the person behind the individual station. A Wang study of managers indicates that the most frequently mentioned and the most perceived problem area is related to the telephone. There seems to be a chronic problem of missed telephone calls combined with an inability to leave an adequate message. Busy phones and unanswered calls amplify the problem.

The study properly identified a managerial desire for tools that would give them better control of their activities:

- Personal time
- Information flow within their environment
- Presentation and display

As for information provision, there have been frequent restart problems during a search through the files for specific pieces of information, though it turned out that active files were generally maintained in the desk and less active files in a secretarial filing cabinet. Furthermore, the systems study must assure that documents to be annotated can be correctly characterized and retrieved. Quite evidently, such activities can be handled in an able manner through computer support.

Another important aspect is that at each level of management the nature of the information elements to be addressed is job-specific. For example, the general manager requires information associated with the whole company operation, whereas the director of finance is mainly concerned with access to financial databases and the tools to promote budgetary control.

In regard to information systems requirements, the Wang study identified six major areas of improvement needed to enhance managerial productivity. The first is voice messaging and mail capability. The second is electronic directories with autodialing and message management. Automated personal administrative support—such as calendar management, to-do lists, and bring-up files—comes close behind. It is followed by voice annotation of documents and messages. The fifth area is word and phrase indexing of documents to assist in the retrieval of filed memoranda and correspondence. The sixth is decision-support applications with color and graphics playing the major role.

Properly directed systems studies can do much to promote management productivity. Microprocessor-supported workstations are most valuable; yet let us never forget that the power of the personal computer is not in the hardware, but in the software. Able software development means the right amount of preparation.

It cannot be repeated too often, throwing money at a problem will not solve it. When it comes to hardware and software, many people overbuy because of their ignorance. The way to a wise productivity investment is through the:

- Able training of the equipment user
- Proper study of the workplace
- Able and timely preparation

Eventually, when we have justified the system, installed it, and made it pay, we will find that we can do so much more with it. The add-on option should always be kept open, but the way to start is from the basics.

Calendar Services

If the new shape of voice communications is the most crucial issue in executive services, it is no less true that its position is being challenged by the developing notion of *time management* and the systems needed to support it. Computer-based calendar services are part of time management; but to appreciate their reason for being, we first have to talk of time measurement and the tools at our disposal.

The two most precious assets a manager has are knowhow and time. The cutting edge of technology and a changing marketplace see to it that knowhow gets obsolete with an astonishing speed. Nothing short of lifelong learning can remedy the situation, and we are going to return to this subject. Efficient time usage is definitely dependent on time planning, time keeping, and the associated time-reporting functions.

Thousands of years ago it was enough to mark the passage of time with the shadows cast by the rays of the sun. In ancient China, people kept track of time by burning a knotted rope; a unit was the travel of the fire from one knot to the next. Then there were the hourglass, which indicated time by the flow of sand, and the waterclock, which operated on the same principle.

Timekeeping became more precise with the advent of the mechanical timepiece, the exact origin of which is unknown. In the fourteenth century, a time-measuring device was called the horologium. The word "clock" originally meant *bell,* and it was first applied to the huge mechanical time indicators installed in bell towers during the Middle Ages. In their early days, clockworks were heavy, cumbersome devices equipped with only one hand, which indicated the nearest quarter hour. A series of inventions in the seventeenth century increased the accuracy and reduced the bulk and weight of clockworks.

Galileo Galilei made a major contribution to timekeeping when he discovered that the period of the swing of a pendulum is constant. Christian Huygens found a way to regulate a clock by employing the Galilean principle of the pendulum. Robert Hooke invented an escapement, which made it possible to use the pendulum with a small arc of oscillation. John Harrison followed suit by developing a means of compensating for the variation in length of a pendulum with change in temperature.

Watches were not produced until coiled springs emerged as a source of power, and then a series of innovations were needed to increase their accuracy. Minute hands in addition to the hour hand were innovations in the seventeenth century. Jeweled bearings were first used by Swiss watchmakers in the eighteenth century.

In the succeeding centuries, before machine-made parts became industrially feasible, craftsmanship of high order was the norm. The Netherlands, Germany, and Switzerland produced craftsmen noted for their aesthetic eye and keen knowledge of mechanics. From that time on, the clock had a dual role in the household: It was both functional and decorative.

Two American horologists, Aaron Dennison and Edward Howard, invented and perfected automatic production machinery in the 1850s. New designs reduced the number of parts required. The 1960s saw the development of the electronic watch, which made use of an electric circuit to ensure accuracy. A tuning fork replaced the conventional balance wheel.

But the real American contribution to the use of time comes by way of applying, through high technology, the proverb: "Time is money." Computers and communications do improve executive and professional productivity. This can best be exemplified by calendar services, including:

- Personal commitments
- Scheduling and rescheduling of group meetings
- Timetables
- Reminders
- Prompts
- Action reports

This calls for a new image of keeping time. A prerequisite for its effective implementation is a closer integration of managerial and professional functions with the computer supports on which they should rest. Note that, although technological advances polished the art of making timepieces, the functional aspect of the clock remained unchanged for centuries. Change has come about with the incorporation of timekeeping functions with DP/WP tools. It takes the form of automatic calendar maintenance, prompting, reminding, the keeping of milestones, and the control of timetables.

Further improvements in the timekeeping function can come about as we study managerial behavior and gather information on what managers do on a day-to-day basis: how they spend their time and the business functions they perform. We know that a manager's day is fragmented and that a great deal of managerial time is spent with other people. A manager is frequently interrupted yet appears to encourage that kind of work environment and prefer face-to-face communication (through meetings or over the telephone) to written information.

To manage time to best advantage, we must familiarize ourselves with the problem of how to increase managerial productivity by identifying what managers currently do in the office, describing the behavior of managers, and identifying the impediments that managers encounter. Only then can we begin to define technological solutions to the problem of managing time. Such research, properly applied, can help clarify the key relations between managerial behavior, advanced computer technology, and the keeping of time. The interrelations must be understood if we are to achieve offices in which people, facilities, and technology are effectively integrated and thereby improve our ability to face the productivity challenge.

Calendar service is one of the tools which can help us to manage time. Through microprocessor-based calendar software, the user can perform time management tasks including the creating, storing, viewing, and printing of a personalized calendar for a day, week, or month of the year. It is also possible to ascertain convenient times for meetings, to schedule routine activities, and to update a list of things to do.

Busy executives can be provided with an electronic notebook in which to store information in free format related to topics of their choice. Notes can be extracted by means of any word or words in the note, enriched with selected notebook entries, and presented on video or printed instantly by pressing a key.

In their daily business, most managers use a desk-size calendar with a view of either a week or a month, whereas their secretaries employ either a monthly or daily calendar. The most common items recorded on the managers' calendars are the time and place of meetings, the purposes of the meetings, and who called the meetings. Some managers also record upcoming events, the things they must do on a particular day, visitors expected, notes regarding the travels of people in their group, and their own travel agenda. In a Wang study, one-half of the managers interviewed kept their calendars on their desks; one-half carried their calendars with them; and about one-half of the whole group said they were willing to let others know when they had open or scheduled time.

Both the managers and their secretaries updated the managers' calendar an average of three times a day. On the average, the managers consulted their calendar approximately 11 times a day. Not every manager was satisfied with his choice of calendar.

Some managers had problems finding calendars that met all their needs: Their calendars were either too large to carry around or too small to contain all the information they wanted to include. Problems occurred if the managers wanted to make appointments and did not have their calendars with them. The calendar systems which the managers were using did not remind them of appointments, so they would have to remember to look at them frequently.

Communication problems between managers and their secretaries occasionally resulted in double scheduling of the managers' time, since in most cases both manager and secretary could make appointments for the manager. Commitments that needed follow-up were made, and coordination was an important subject.

The managers interviewed in the Wang study encountered some major impediments when using *to-do* lists: Manually writing the list was a time-consuming and cumbersome task, particularly when the information was transferred from one slip of paper to another. There was also a tendency to lose the list. Dictating the information was easier and faster, but the list had to be transcribed before it could be used. Furthermore, the managers did not always remember to look at their lists, and there was no way the current system could provide prompting and reminders.

Projecting to computer-based calendar services, all managers interviewed wanted to see blocks of time in their calendars to determine quickly whether their time was filled or open. They also wanted the facility of knowing, at quick glance, the number of meetings they had scheduled in the morning and afternoon and if they had time to schedule something else. There was a good deal of similarity among the managers as to what they actually recorded on their calendars:

- Information about meetings and travel
- Upcoming events
- Expected visitors

Interviewed managers suggested that a computer-based calendar should allow them to record more information:

- To-do lists
- Notes of things to bring to meetings
- Things they wanted to talk about to visitors
- Travel itineraries

A time-consuming function that was performed by all but one of the managers' secretaries was to transfer information between the manager's calendar and secretary's calendar so that both calendars contained the same information. This was usually done manually on a routine

basis every morning or evening. Computer-based solutions help cancel
the need for manual approaches.

Clearly, calendar automation should enhance the ability to maintain
absolute privacy of individual calendars and to-do lists without reducing the overall effectiveness of the system. It should also facilitate the
concurrent use of multiple, closely coupled activity management and
communication tools such as calendar and other databases.

Finally, another issue to be carefully weighed is a facility to allow
secretaries to access and maintain their managers' calendars so that
both the manager and his secretary always have the most up-to-date
version. The same thing is true of information recorded on previous
calendars and to-do lists and reminders of time-dependent events.

Management Graphics

The changing direction of system studies can be nicely exemplified by
the new audio services and the calendar facilities. For nearly three
decades, the lion's share of computing has been taken by applications
that support routine organizational operations: payroll, simple record
keeping, billing, and engineering studies of the number-crunching variety. Few computational resources have been devoted to supporting
the ability of managers to schedule, allocate, and control their resources. The rarest of all applications were long-range planning and
policy analysis.

But times are changing. Computer graphics have been used for two
decades in engineering, architecture, and military applications known
as computer-aided design (CAD). Recently, both financial institutions
and industrial companies have begun to use computer-generated
graphs and charts as presentation tools in boardroom settings. Management is at the threshold of integrating computer graphics into decision
making.

Over the next few years, the growth of computer graphics in business
applications may outstrip all other uses. It is estimated that, by 1986,
one out of every two graphics terminals shipped will be for business
applications, compared with only one in eight or ten today.

The reasons for this trend are almost self-evident. Image, text, data,
and methods that help focus attention and evaluate choices improve the
technical performance of a decision maker. Computer-based information systems can help managers act more rationally by enabling them
to compensate for the weaknesses in their abilities to select and remember information.

Propelled by an accumulated knowhow, computers and communications can help organize and filter larger volumes of information. Simu-

lation enables management to consider a wider variety of complex dynamic situations. Optimization helps in scheduling and in the allocation of resources. In this sense, computer-based systems become very helpful instruments.

A word of caution is necessary: Computers and communications can be so much more effective if the proper organizational work has been done, authority and responsibility are streamlined (Fig. 2-1), and the lines to decision making are opened. Clogged communication networks and monolithic organizational structures lead to corporate entities that exist only by a kind of separation between official ritual and reality, and there office automation (OA) systems can make no contribution to efficiency.

The key word in the use of computers and communications for decision support services (DSS) is "visibility." The success of the Polaris missile project has often been attributed to the use of PERT scheduling, which was developed for it. What is often forgotten is that PERT was *not* used to manage the Polaris schedule and costs. The admiral in charge of Polaris development was not particularly concerned with how PERT was used. His emphasis was on PERT's being visible to all parties who contributed to the project.

In any managerial activity the raw data are interesting, but they are ineffective if they are not aggregated and presented in an action-oriented manner. They rarely are. Most frequently paperbound management reports do not provide adequate, valid answers to the questions which are critical to management decision. To the manager who must make decisions on multifunctional data, the value of information is directly proportional to the ease of understanding and the comprehensive presentation.

Therein lies the basic reason why the problem of getting the information out of the computer in a usable form and disseminating it to the appropriate decision makers has become a cornerstone in systems design. At the same time, the trend toward integrated information processing—which includes data processing, word processing, and image handling—presents a good opportunity for efficient solution.

The fundamental premise is that by tapping the brain's vast capacity for nonverbal perception, computer graphics can communicate as much as two orders of magnitude more effectively than statistical printouts alone. Trying to interpret the rows and columns of statistical data is an inexcusable waste of managerial time. People who have to base top-level decisions on valid information are those who have the least time to manipulate raw statistics.

Sending raw statistics to middle management for interpretation and summarizing creates delays. But it can also garble, deemphasize, or

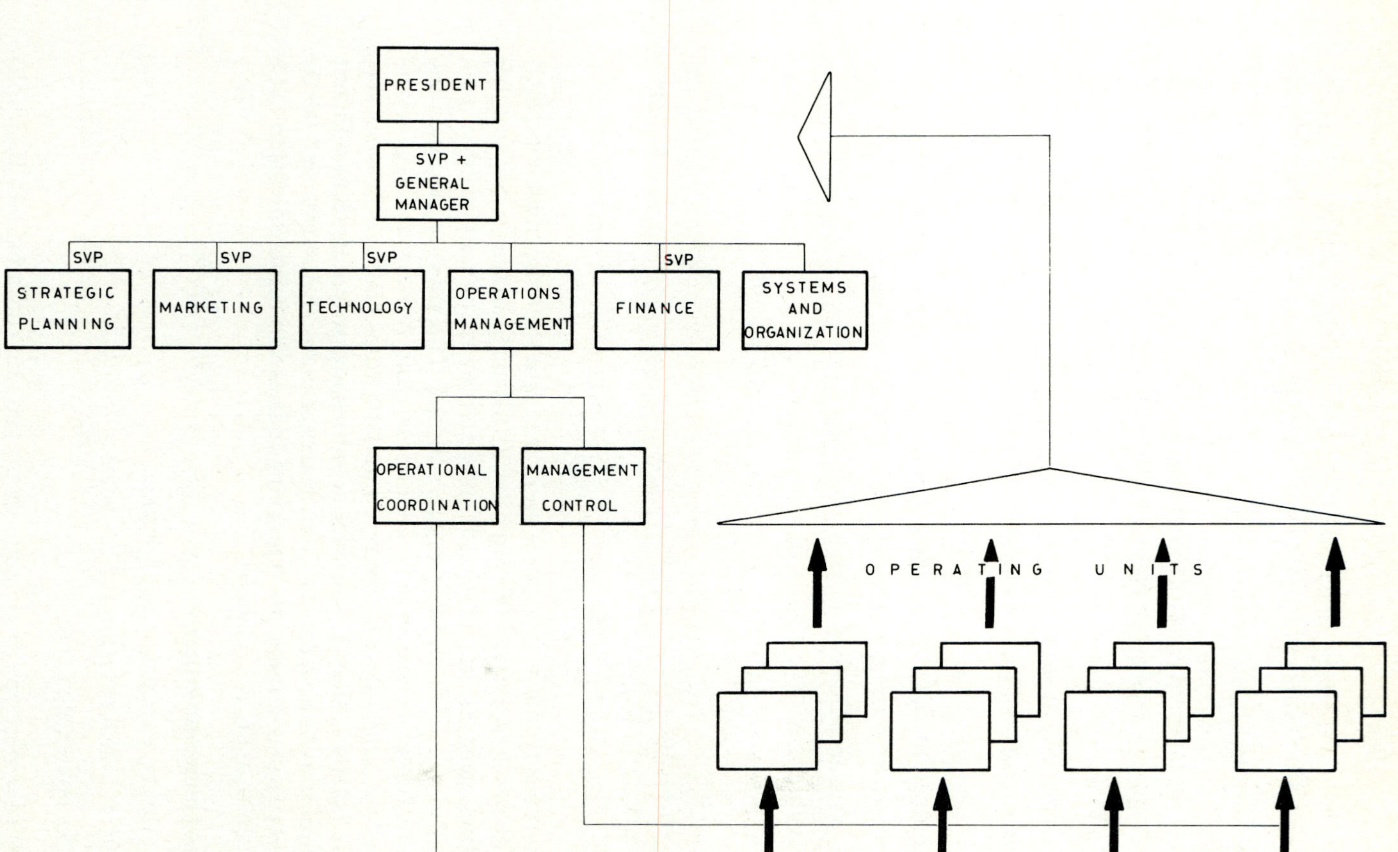

Figure 2-1 Streamlining authority and responsibility is a prerequisite not only to a valid organizational structure but also to effective implementation of computers and communications.

cause to be completely overlooked information that is the key to decision. Delays and misinterpretation are two of the most common reasons for corporate failures, and they can be found in the background of the fifteen topmost reasons why a business gets into trouble:

1. Lack of forecasting and planning
2. Changes in the marketplace
3. Changes in technology
4. Product obsolescence
5. Uncontrolled increases in the cost of production
6. Limited financial resources
7. Increase in the cost of debt
8. Growth of business beyond skills of management
9. Decreased efficiency of the sales network
10. Development of location disadvantages
11. Management short of guts to cut out dead wood
12. Internal conflicts
13. Obsolescence in knowhow
14. Inadequate control systems
15. Tarnished image in the marketplace

Recent studies of visual perception indicate that the old saying that one picture is worth a thousand words actually undervalues graphic communication. If information is conveyed by using only printed messages, the visual channel is underutilized: The total capacity of the visual channel—the eyes, optic nerves, and visual centers of the brain—is estimated at between 30 and 40 mbps (million bits per second). It is, then, easy to see why graphics are so much more efficient than alphanumeric presentation. Furthermore, if pictures are better than words and numbers alone, color makes the communication even more comprehensive. Color

- Adds clarity and impact
- Highlights, separates, and defines
- Associates information on the screen
- Points out exceptions

As a result, relations are vividly displayed and complex data is simplified. Color adds immediacy, improves visualization, and speeds identification. Since color is more interesting and visually demanding, it increases operator attention. This cuts response time, human error, and fatigue and improves man-information communication.

For managers accustomed to lose time over stacks of computer printouts, graphics are a welcome relief. A General Motors study suggests that the use of computer-produced graphs instead of long computer printouts has trimmed to as little as 20 min meetings that used to run longer than 2h (hours).

At General Mills and Boeing, computers are now converting the endless columns of numbers into colorful charts, graphs, and maps that are helping managers spot trends and make decisions more quickly and efficiently than ever before. At Boeing Commercial Airplane, a division of Boeing, charts that once took as long as 2 weeks to develop are now produced in a few hours. They enable managers to see more up to date data and request new presentations on short notice and explore a lot more alternatives. At the Cadillac division of General Motors, road maps and other graphics, which in the past were drawn up manually, could take as long as 2 years to complete. By the time management got them, they were obsolete. Now, the data gathering might still take months, but the maps themselves are produced in less than a day.

With graphics, managers can absorb three times as much information and in less time. Computerized graphs can forcefully alert managers to potential problems before the problems get out of hand. Also, managers can use graphics terminals directly, instead of requesting charts from their data processing staff. To be sure, few observers expect senior executives to sit at graphics terminals, but the middle manager can look at 20 representations and then decide which 6 he wants to present upstairs.

Furthermore, top managers will be able to play more with what-if models; they can ask the computer to plot the results of more than one business scenario. Financial executives can use the computer-mapping system to project the variation in revenue that would result from various placements of funds. Marketing managers can experiment with the assignment of salesmen.

Yet, because computer graphics is still a new tool, it has to be used with care. We should definitely avoid the repetition of certain failures so common in the past. The first and foremost rule (almost invariably violated in tabular-form presentations) is that read ability should always be supported. That is what Figs. 2-2 and 2-3 are for. The first concerns a banking application; the second can be used to format any time from series, although this specific example has to do with some statistics from *Chemical Abstracts*.

Management graphics is definitely a communications tool: the best available at the management level. Information systems must use graphics to present, for decision support, the contents of the database and thereby pinpoint trends and failures. But graphics presuppose establishing standards to be universally understandable and not misleading.

From the design standpoint, we have to assure that, with the help of a computer graphics system, we can put all management-oriented information into a form which can be understood quickly and simply. Data must be graphed, charted, and trended as the direct output of the computer. And it must be made available online and interactively.

Graphic presentation must reveal trends and give an immediate sense of impact to management. But the use of graphics and color within the perspective of a management information system (MIS) also requires a considerable amount of preparation. Chase Manhattan identified four stages in developing its MIS:

1. Standardization

2. Collection

3. Flow management

4. Presentation

Color graphics come into the picture in stage 4, but in a new and forward-looking project the intended use of text and data should be reflected from stage 1. Standardization concerns all MIS component parts. Financial and performance information, for instance, will have six parts:

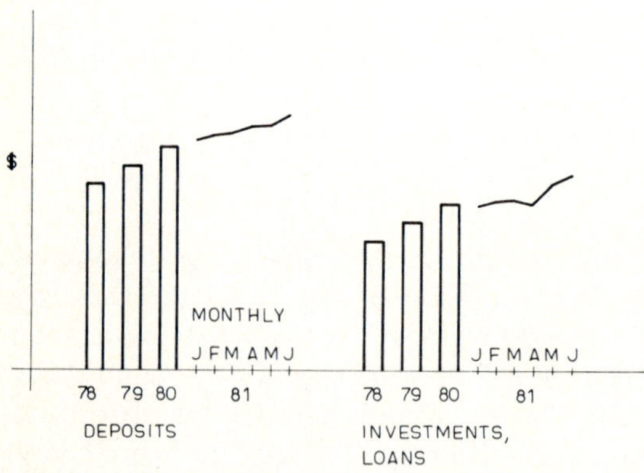

Figure 2-2 Graphical presentation of deposits and loans as a substitute for tabular data to be used by management of a medium-sized bank. The presentation can be by branch office, region, or country or even international.

- Organization
- Customer market segment
- Product segment
- Customer profitability
- Product profitability
- Portfolio

Since every information system will necessarily use some or many of the existing test and data facilities, the characteristics of an MIS will:

- Be based on existing production systems
- Rest, for updating purposes, on valid and timely data capture
- Use a common input discipline
- Use common-denominator information elements
- Operate within a common systems environment

Of the outstanding MIS issues to be addressed, presentation display and integration of activities are the most important.

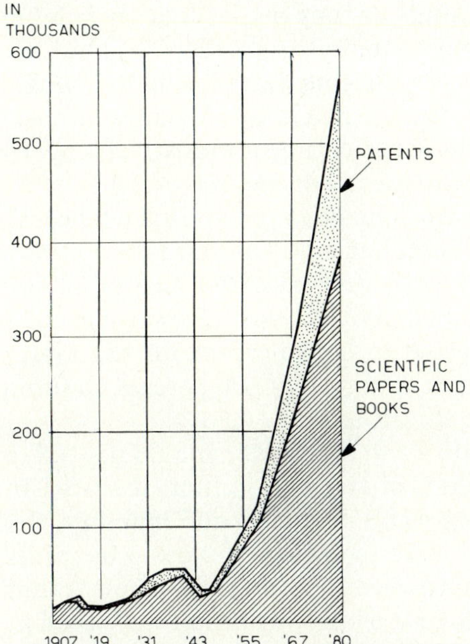

Figure 2-3 Statistics from *Chemical Abstracts*. This is an example of a chart which can be presented effectively in color.

3

Communicating Workstations

"It is a good thing to follow the first law of holes: If you are in one, stop digging." DENNIS HEALEY

If the organization is happy with things as they are—typically, a lot of mainframes working batch and timeshared minis on dispersed but not necessarily coordinated operations—there is no real reason for change. Chances are, however, that the current, often classical computer operations are a hole in which the financial institution or industrial concern found itself because of a lack of strategic planning.

In many companies, computer operations have grown like wild cacti, and that is true of the number of installations. On January 1, 1983, there was in the United States an estimated installed base of 56,000 mainframes, 570,000 minis, and 3,500,000 PC, with growth for 1983 and the next couple of years projected to be 5 percent for the mainframes, 11 to 13 percent for the minis, and nearly 100 percent in units and 50 to 60 percent in money for the PCs.

There is a trend in equipment utilization the reader should not fail to notice. Figure 3-1 demonstrates that, of the $3.4 billion invested in 1982 in locally distributed systems by U.S. business and industry, the largest share (44 percent) went to cluster solutions, followed by minis and small business systems and mainframes. The projection now is that if the same criteria are used in 1986, first of all the market will stand at an impressive $14 billion and second, the largest share will go to PCs and LANs.

How much of this installed base could answer the users' requests in a valid manner? An MIT study made in mid-1980 demonstrated that:

- A full 33 percent of the important activities in a firm had *no* computer support at all, although they could be effectively supported by computers.

- At the same time, only 17 percent of the existing installed base of information systems was *relevant* and *appropriate* to users' important needs.

What is worse, the situation has changed very little since 1980. Only now does the need for a turnaround become a matter for concern.

Part of the hole in which the users and the providers of information find themselves today is resistance to the implementation of newer system concepts—keeping up with what technology can offer. Star-type connections, though valid for about 20 years, are now deadly obsolete. The modern, efficient approach is with PCs and LANs. Yet, although

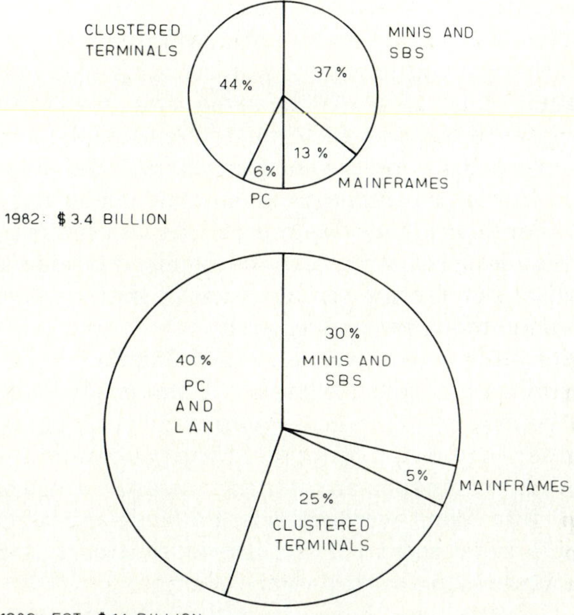

Figure 3-1 The U.S. market for distributed information systems and the money invested. In 1982 the market for locally distributed systems stood at $3.4 billion and the largest share went to clustered terminals. The same market in 1986 may be $14 billion with the largest share held by PCs and LANs.

star-type connections are inefficient, tax the resources of a small mini, increase response time and delay turnaround, they also have friends up the organization who resist their replacement.

It is not at all uncommon to find that computer applications are characterized by *discontinuity* and are aimed at protecting the *power centers* created by the exclusive holding of data rather than helping the end users in their work. How often we fail to appreciate that information, like knowhow, grows as we share it. Technically as well as financially there are valid reasons to have, by the late 1980s, at least one WS per professional worker.

The real issue is not personal computers, but *personal computing*. Clever design, packaging, merchandizing, and publicity helps a product find its niche in the market and eventually become a success. The same principles should prevail in information systems design and implementation.

Management Productivity

A change in management attitude toward the use of high technology in daily work should be part and parcel of the dramatic shift under way in our industrial society. The emerging science-driven economy sees to it that the transition will not be painless. At the same time, this is an inevitable development in an economy which must now compete on a global scale: United States companies cannot pay the $20 per hour demanded by our skilled work force to create products that can be made in the Pacific Basin at $4 per hour labor costs and still remain competitive. We have to generate jobs that create high-value-added products. Computers and communications are the best products of that type we can find, as clear-eyed companies now understand.

At United Technologies, 1000 executives who earn $50,000 a year and up, take 3-day courses on personal computers. Upon graduation, each is issued a personal computer with video, printer, and other accessories to use in any manner that seems suitable. In the background of this policy is a major switch in management attitude toward the use of computers as individual, intelligent workstations. So far, computers, although commonplace at lower corporate levels, are not routinely used in executive suites. The underlying attitude must change.

More useful to corporate executives, however, will be the next generation of multifunctional, communicating workstations. They will include not only computing power but also word processing, file access, telephone facilities, and dictating machines that respond to spoken commands. The workstations will be linked through local area networks to other stations, minis, and mainframes so that managers can share data and exchange messages. They would be able to provide a wide range of services including electronic mail, financial control, strategic analysis, and graphics preparation at the best cost.

For example, Manufacturers Hanover Trust, seeking ways to streamline its channels of internal communications, found the electronic mail system it installed in 1980 to be of great value. A management estimate indicates that about 19 percent of the user base is senior management, 28 percent is professional, and 53 percent is middle management. Currently there are some 2400 users sharing over 1000 terminals on the system. The biggest savings come from the elimination of phone calls, followed by the elimination of many paper memos and meetings.

It is hard to measure the productivity added by using wide-ranging electronic mail, but many users have estimated, through surveys, that the system saves 30 to 40 min/day (minutes per day). Multiplying that by the number of employees and by the average pay, one derives an estimated savings of millions per year in time alone, for a 2000- to 3000-people company.

Manufacturers Hanover Trust has also been involved with networking. In Manhattan, where its headquarters is located, the bank has a metropolitan area network based on cable television (CATV) technology. Several locations in New York City have been connected with a channel on the cable network, and the bank's corporate headquarters employs a broadband LAN. The expected saving to the bank, by replacing telephone company charges, was expected to be $1 million in 1983.

Still, management's determination, though absolutely necessary, is not enough. The analysts must take care of detail, and they should assure that the assets provided by technology are being exploited to everyone's advantage. Workstation design is an example. Till now the user has had to sit at the computer keyboard and type in arcane and hard-to-remember commands to make the computer work. Even on a personal computer, every time the user wanted to change from one job to another, it was necessary to change program disks. What is more, the information typed into one program had to be typed in again if it was needed in another procedure. Yet developments in computers and communications are such that, with the right WS implementation, the user can interact through easy media: touch-sensitive screens, tablets, joysticks, mouse, and eventually voice to feed in data and get the machine to work.

WS software lets the user do jobs such as graphing, word processing, calculating, drawing, list creation, and project scheduling all on the same stations. In a way totally transparent to the end user, jobs can be brought to the top of a pile or placed underneath for future reference. Data can easily be exchanged among jobs.

The user can tell the WS what to do by pointing to one of the commands listed across the top of the screen, eliminating the need for keyboard. The pointer arrow can be moved by a mouse or a joystick. A click of the button activates the chosen command, and a set of follow-up instructions appears.

Housekeeping chores can be effectively represented to the end user by small pictures on the screen; for example, data can be thrown away when moved into the wastebasket or be saved on a clipboard. Typically, such workstations will:

1. Be designed for managers and professionals

2. Have a heavy graphics orientation, be menu-driven, and use an easy way to manipulate the cursor

3. Simplify the handling of the screen, as by pressing a button to display commands, open or close a file, or move a word or information element

4. Make it possible to create and maintain any type of lists in a personal database—the WS *microfile*

5. Be supported by integrated software so that the user can *cut* what has been created by the use of one package and make it available to another.

Presentation and display studies should not only consider the positive aspects but also focus on the impediments to productivity, as perceived by the managers in their day-to-day work environment. Impediments include issues fostered by organizational and environmental constraints and other factors that are behavioral in nature. Impediments may exist within five major areas:

- Oral communications

- Personal administrative support

- Text and data capture

- Filing and retrieval

- Output creation and presentation

To resolve the prevailing miscommunication problems, the study should aim at drawing a broad profile of the organization and also selecting the factors to be subjected to in-depth follow-ups.

A study done by Wang to establish the way managers communicate with the systems under their control revealed that more than 80 percent of the managers shared secretarial support with three or more people. Over 60 percent of the managers reported that they spend less than one-half of their time at their desks.

Two-thirds of the managers' communication was verbal. Interestingly enough, more than three-quarters of the respondents have at some time had computer terminal experience. Nearly three out of four managers maintained their own calendars exclusively.

- Half of the managers studied indicated that they carry their calendars with them.

- One-third of the responding managers frequently felt they received insufficient notification of meetings.

- Three-quarters of the managers indicated that they were frequently interrupted while reading in the office.

The respondents' ranking of preferred reading locations is listed in Table 3-1: The weighted rank represents total points accumulated based on the assignment of three points for first choice, two points for second choice, and one point for third choice.

Another interesting statistic is that the majority of managers typically spent less than 35 percent of their time creating documents. Their ranking of preferred document creation locations is shown in Table 3-2. A frequently cited problem was the managers' inability to categorize documents uniquely, indicating a need for a highly flexible filing and retrieval capability.

Justifying the Cost of the Executive Workstation

In implementing office automation and decision support systems, we must start from the areas carrying the greater weight: An estimated 75 percent of office costs is attributable to the managerial and professional personnel. At the same time, most of the information elements necessary within an organization are particularly valuable in one location, the one at which work is done:

- An estimated 75 to 90 percent of the data used in a local environment is generated within that environment; the balance comes from the outside.

Table 3-1   Reading Locations Preferred by Managers

Location	Rank	Weighted rank
Office	1	66
Home	2	44
Commuting	3	34

- Similarly, some 75 to 90 percent of the information produced within a local environment will be used within that environment and the balance sent to the outside.

As a result, the majority of communications are local: The largest proportion of information used within a department has limited value outside it. The greatest part of information exchange takes place within a relatively limited organizational area, but a substantial share of local processing is *interactive* in nature and *bursty*. A softcopy (video) must be filled with large bursts of data at high speeds. A similar statement can be made about file transfers from one database to another. Whether transmitting text and data files or high-resolution graphics, the system requires lots of channel capacity.

In handling high-volume traffic, message routing and network control should introduce as little delay as possible; it should provide speedy access to the database for all nodes and users. The network architecture should be chosen to support:

- Packet switching protocol (preferably international standard; applications-independent)

- Network control centers (online diagnostics, quality histories, assistance for self-maintenance)

At the same time, system design should stress databases and make them:

- Integrated and thus open to partitioning, distribution, layering, and expansion

- Able to handle text and data; supported by dictionaries

- Protected through security mechanisms for authorization and authentication

Such developments call for new perspectives and a new orientation. In the latter, stress should be put on solving the problems which currently prevail with office automation. Among them, three issues are particularly worrisome:

Table 3-2   Document Creation Locations Preferred by Managers

Location	Rank	Weighted rank
Office	1	78
Home	2	44
Commuting	3	25

1. *Concept and strategy are missing.*   Not all companies have a clear concept of OA, and therefore they cannot formulate a *competitive* strategy. They are handicapped in setting objectives and making budgetary evaluations.

2. *Divided procurement.*   Procurement responsibility is often split among MIS/DP, office equipment (typewriters, copiers), and telephone services. No clear-cut lines of responsibility exist, nor is there the needed coordination. A valid division will give MIS/DP the responsibility for any equipment which has a microprocessor, RAM, disk, and data communication capacity and is programmable.

3. *Lack of the needed knowhow.*   Specialists with OA skills are in very short supply. Furthermore, a consistent effort must be made to convert to OA not only the specialists but also the users.

The best knowhow must be applied not only in technical systems design but also in cost optimization. Table 3-3 shows why. The cost of the workstation is steadily dropping. In 9 years the dollar was inflated by more than 80 percent. If steady prices had prevailed, that would have taken the cost of a workstation to $25,200 in 1985. Instead, the expected cost of a fairly complete WS in 1985 will be about $3000. That is an order-of-magnitude difference.

The IBM personal computer, for example, currently provides processing capability at least equivalent to that of the System/360, Model 30, which was the most popular commercial data processing system of the late 1960s. Yet personal computers cost less than $5000, whereas the cost of the IBM Model 30 ran into the hundreds of thousands of dollars in the late 1960s.

In the industry the trend is discussed in terms of *price/performance:* The technology provides the means to drive prices down while holding performance constant or to push performance and functionality up while maintaining prices constant. The choice is to minimize price or maximize performance. Generally, a rate of 20 percent has been a reasonable price/performance trend for many products. At 20 percent

Table 3-3   Projected WS Market in the United States

Year	Complete WS	Avg. profess. yearly salary	Ratio, %	Units, millions	Market, billions
1977	$14,000	$30,000	47	0.10	$ 1.4
1980	12,000	42,000	29	0.20	2.4
1983	6,000	55,000	11	0.80	4.8
1986	3,000	70,000	4.3	4.0	12.0

annually, the price of today's $5000 system can be cut to a tenth, or $500, in only 10 years while maintaining constant performance.

At the same time the typical cost of an executive or professional position is projected to move to the $70,000 level in 1986 (from $30,000 in 1977). Some 60 percent of the difference is compensation for inflation. The balance is real gain due to the scarcity of qualified personnel.

These facts of life lead to two interesting figures to be interpreted as leading tendencies. Figure 3-2 contrasts the salary curve to the cost of the workstation. The ratio between the cost of a workstation and a yearly executive salary is most important. It moves from 47 percent in 1977 to a projected 4.3 percent in 1986. No wonder the demand for executive workstations may skyrocket. Figure 3-3 makes this point: Not only the cutting edge of technology but also the growing demand for management support systems is expected to fuel an unprecedented growth in the demand for workstations.

In the last analysis, the question is not how much the executive or professional workstation will cost but how much profit the investment

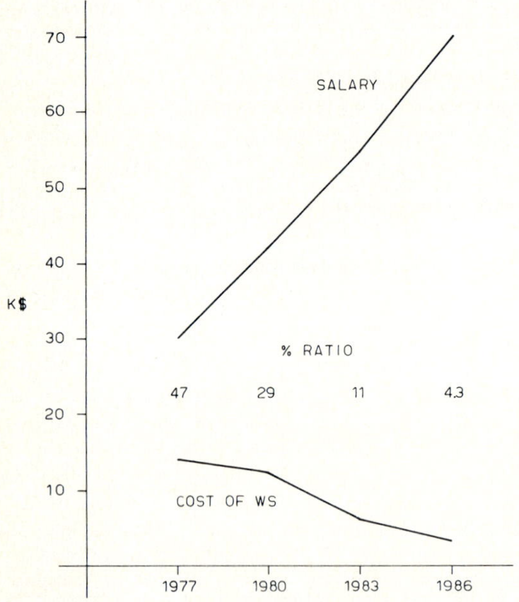

Figure 3-2 Cost trends in office work. Executive salaries have steadily increased and are expected to continue to do so because of the scarcity of qualified personnel. On the other hand, the cost of WS is continually dropping. The ratio between the two cost items has fallen from 47 to an estimated 4.3 percent.

will bring. Thirty years ago, my late Professor Neil Jacoby taught me: "A banker will be happy to spend a dollar if he can make two dollars benefit out of it." Executives who have managed to master their workstations have commented that their productivity has increased between 10 and 20 percent. With the proper preparation, we can top that figure.

Providing the Text and Data Transport Facility

Computers are no longer the simple DP propositions they used to be. They are vital parts of a broader communications and reporting system linking the "person with the problem" to the information through a protected and updated access mechanism. This must be done:

- Interactively
- In a user-friendly manner
- Without excessive training of the user
- By applying new but known technologies

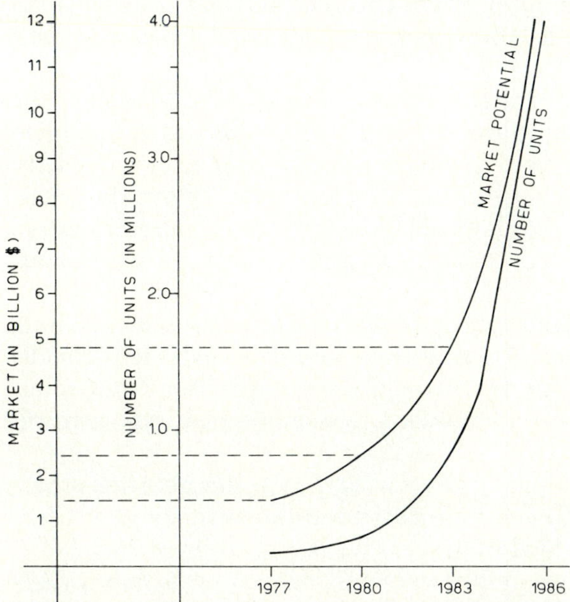

Figure 3-3 Growth in work stations. The growing demand for management support systems, fueled by competitive needs and technological advances, is expected to cause an exponential growth in WS demand.

For the change to be effective, we must alter the image about information systems both among the users (management, day by day) and the specialists, which involves a lifelong training task. Within this perspective, one of the key challenges is the proper allocation of people, money, and means (technology, organization, and products). Another is the fulfillment of the prerequisites for going online:

1. A network—both local and long-haul—reaching easily accessible points

2. A terminal facility for input and output (preferably softcopy)

3. Rules of conduct (protocols)

4. Supporting software

With reference to these four key factors, compatibility has emerged as a major issue in the growth of the DP/DB/DC base of an organization. Innovation has to be balanced by continuing support of current services till the effective replacement of those services by the new ones.

Sharing resources, most particularly the software investment, is also of vital importance. A New York bank is developing a network in which processing power will be shared so that programs that would otherwise be too large for a microcomputer can be run. The network is intended to impose no essential limit on the size of programs and amount of data that can be processed.

The local area network consists of five PCs and a 33-MB disk of the 20-odd vendors evaluated, 18 were eliminated because their products were not yet available or were too new or because the excess capacity they required cost too much. Compatibility, however, delays the wider spread of the network. The bank would like to see other microcomputers besides Apple supported, and software assistance for multiple operating systems supplied.

Bank management intends to put local area networks and PCs in its branches to automate much of its clerical work. The LANs will also be linked to the central mainframes through a communications controller, and considerable savings from reduced telecommunications requirements are expected.

Within a carefully outlined systems study, once the network structure has been defined, the communications links must be identified. They will typically be formed in several ways: privately owned cable, common carrier leased lines, public value-added networks, public dial-up lines, land-based microwave, and satellite communications. The user must be familiar with both the characteristics and the economies of these links before making a proper selection.

Among the common compatibility problems that must be resolved at the design stage are data coding, communications protocols, and electrical interfacing. Mainframes will be around to run big databases and act as communications switches. Yet most mainframes manipulate data that has been encoded by using EBCDIC, whereas nearly all mini and micro computers use data encoded by using ASCII.

Protocol differences are also a problem for the network designer. The protocol is involved in framing bits (or bytes), controlling errors, numbering messages, assuring transparency, managing lines, timing, doing start-up, and resolving conflicts or other transmission problems. There are various communications protocols, but no universally accepted standard exists (although HDLC and X.25 have been specified by international bodies).

The present trend is to packet switching, and here again technology has helped greatly. In the mid-1970s, packet switching had a nearly 28 percent overhead. By 1982 the implementation of virtual circuit brought the overhead down to 8 percent, with datagram standing at the 12 percent level. (This information was given in a meeting with the communications specialists of Metropolitan Life Insurance in October 1982, in New York City.)

The statements about protocols are valid all the way into the applications environment. Whenever an online implementation is projected, the establishment of universally applicable input/output conventions is a fundamental prerequisite. How far formatting should go depends on the population and the importance of the interactions. If the population is large, then we should:

- Standardize nearly all elements

- Augment the information content by computer (central information file, data dictionary)

- Usually report by exception, but provide access to detailed information elements

- Generate use statistics to plan for economies of mass both for text and data transfer and for databasing.

If just a selected, limited population is involved, then we should:

- Standardize at least what has been done in DP/WP

- Provide for integration of other classical means, such as paper via indexing (directory), and relate those means to the online database

- Look into microform capabilities or, better yet, optical disks

Still other problems in network design involve access control, network control, customizing nodes to the application, evaluating risks of program downline loading, and establishing communications software sharing between hosts and nodes. As those problems demonstrate, choosing a distributed network architecture that will meet all of a user's expectations is not an easy task. The task is made even more complex by communications and computer suppliers lacking standard definitions and a universal network architecture philosophy.

A number of critical questions should be answered. One of them is how long to keep a journal. The answer depends on the goal.

1. If it is *restart/recovery,* we must keep the online trace as long as we have the requirement. Operational needs overlap with one another; a "rolling journal" can handle them, but we usually change the "live journal" once a day.

2. If *security* is the consideration, it poses a long-range requirement, but it can be managed offline.

3. If *legal requirements* exist, they must be met. If they do not, then 1 year is the minimum and 5 or 6 years is better for keeping the journal trace.

Also, a concrete effort should be made to emulate and benchmark the operation of the online, interactive environment under development to avoid unpleasant surprises after it becomes operational.

The last few observations point to the fact that a network design product typically involves lots of software. Software has a growing impact in the field of information transfer. It is replacing some hardware functions (and vice versa), assuring network control, and helping to bring about electronic mail. We all know that software is the most expensive component in many data networks because efficient approaches require the services of personnel whose functions are not yet fully automated and whose expertise is rather scarce and costly. Further complications are the proliferation of advanced control features which must be programmed and the often-encountered lack of good packages in this field.

An Insurance Company Builds Its Network

In early 1979 Metropolitan Life Insurance, of New York, started designing its MIS-oriented SONIC (sales office) network on minis and personal computers. IBM S/1, HIS L6, and TI 990 are three types of minis used in this communications aggregate, which links eight head offices with regions and districts. (The SONIC experiment is at the district office level, which has seven to eight people.) The applications include:

- Management information
- Insurance illustration (for applicants)
- Data entry (into the local machine and onward to a mainframe)

With the IBM PC coming into the picture, management is taking some applications from the minis and transferring them to the microcomputer. (As a senior executive remarked, the IBM PC use put an end to the IBM 8100 experiment which had been ordered.) More of that is to come. Presently with the minis, as management of the insurance company pointed out, clerks use some functions and not others because they are not user-friendly. With each PC dedicated to one person, the change is radical.

1. *Data entry at the local level has been significantly simplified.* Specially designed software and input devices help the clerks enter the applications for insurance and subsequently forward the file to the mainframe for processing.

2. *Sales agents use the PC to do prospecting.* This integrated application involves client handling, report generation, and the evaluation of P/L. The mini does all that also, but it is too slow.

3. *New managerial functions have been implemented.* They include MIS, inquiry, file verification, commission, evaluation, premiums, electronic mail, and a cash quote system. Electronic mail is now used for data communication as a substitute for voice telephoning.

4. *The application base is being refined by advancing PC installation to the agent level.* This needs emphasis. With minis it would have been too costly to expand the network of intelligent machines to the agent level. The PC makes this expansion not only feasible but also mandatory.

Metropolitan Life Insurance is now actively planning for the integration of voice, data, text, and image so that it will not have to change its network design with the evolution of the applications environment. Early estimates indicate that, by integrating voice, text, and data, the resulting load will most likely be made 88 percent voice and 12 percent text and data. The project calls for handling voice, text and data, and image at high speed on the same pipe; 6.2 Mbps (megabits per second). The network will cover all of the United States with voice being given preference.

Economies of scale are behind this drive. According to projections made by the specialists, the use of high technology in voice handling will save the firm some $75 million in 10 years. The specialists are also projecting a complete end-to-end effort to avoid the use of gateways, which are expensive propositions involving delays and possible errors. (Gateways are software-intensive, and that drives their costs up.)

The example demonstrates the great importance of technical and economic studies. There is no way to avoid the fundamentals in a new information systems design, and that calls for rituals involving seven basic steps:

1. *Specification.* What is the projected service? The expected performance? The budgeted cost?

2. *Design.* How will we organize, structure, and deliver the service?

3. *Development to production.* What is the timetable for proceeding with the actual delivery of the service to the client as specified in the design?

4. *Testing.* What kind of assurances do we provide that actual delivery is within the design specifications? And that performance evaluation has given good results?

5. *Implementation.* Have we outlined the procedural steps for end-user-level application? For handholding? For seeing through the teething troubles?

6. *Maintenance.* What is our plan for preventive maintenance diagnostics, tuning, and self-assisted repair of breakdowns?

7. *Feedback.* The output of each stage is the input to the next one, and a system feedback is also necessary. How do we plan for it?

There are other vital questions: Are we really starting the information system at point of origin? Do we deliver information strictly at the *end user* level? Is this done in a simple, easy, prompting, and forgiving manner? How much does this solution aid mental productivity? Clerical productivity? How cost-effective is the resulting system?

Basic questions about benefits include these: Does the overall design of our facilities affect, in any substantial way, the productivity and quality of the work life of our people? The managers? The professionals? The clerks? Which specific aspects of our environment cause these effects and in which ways? What are these effects worth, and how do they compare to the costs of providing more effective operations?

Efficient, factual and documented answers call for diagnostic evaluations of the workplace(s) and their effects on productivity. The evaluations should be followed by the development of programs, guidelines, standards, and measures of the effects of changes. Only then can the economic analysis of the cost of problem solution and the benefits of change be meaningful.

Electronic Mail and Videotex

*It is impossible to make anything foolproof,
because fools are so ingenious.*
ONE OF MURPHY'S LAWS

Part of the confusion about videotex, Viewdata, Teletext, and Teletex is due to terminology. The process originally launched in England (which we will cover in this chapter) was called *Viewdata*. It supported the exchange of messages in a package format, and it involved the use of telephone lines and a home television set.

Two other message exchange processes developed almost parallel to it. The one, *electronic mail,* has been a point-to-point service particularly promoted in the United States. Early offerings, like Telemail, involved nonintelligent terminals. The advent of personal computers made it very easy to use low-cost intelligent terminals through the acquisition of packages like Microlink (a $90 bargain).

The second parallel development, *Teletext,* is a broadcasting service. Its object is the transmission of text and data frames employing part of the unused capacity in a television channel. It is a service of significant importance not only for communicating information but also for supporting handicapped people who cannot listen to speech.

Teletex is the Consultative Committee for International Telephone & Telegraph (CCITT) term for electronic mail; it can easily be confused with Teletext. The CCITT term for Teletext is *Broadcasting Videotex,* and that for Viewdata is *Interactive Videotex.* Since CCITT is one of the international bodies whose function is coordination and normalization, it would be proper to lean toward using the terms it proposes. Unfortunately, 3 years after their proposal, those terms still had not filtered down the line:

- In Europe, where CCITT has its main strength, several broadcasting stations use "Teletext" rather than "Broadcasting Videotex," and "electronic mail" has in no way been replaced by "Teletex."

- In the United States, "electronic mail" is firmly entrenched in the terminology, Time Inc. is coming out with "Teletext," and Compu-Serve is selling a "videotex" service.

In view, then, of this de facto standardization, we will use the terminology electronic mail (not Teletex); videotex (and not Interactive videotex or Viewdata); Teletext (not Broadcasting Videotex). From this point on, the terms will be lowercase, not capitalized. VDB will be used to indicate "viewdatabase."

The preceding discussion should also be taken as a warning that not only is terminology not yet standardized but also such services are in steady evolution. Some of the services are substituting for others: The CompuServe videotex service includes electronic mail, and at least one LAN architecture supports electronic mail and videotex in both an alternative and a complementary sense.

This chapter is introductory. Now that the dynamics of the new interactive message services (the missionary work) have been discussed and the supports the busy executive can obtain from them have been presented, it is time to review in a comprehensive manner what these services are.

What Is Electronic Mail?

Electronic mail can be implemented in three modes. Most elementary is by way of hardcopy devices attached to a network which supports message-switching exchanges. This is an extension and modernization of the classical telex solution using telephone lines rather than a dedicated message-switching network.

An alternative approach is to use faster input/output (I/O) devices than the traditional teleprinter and a formal discipline of exchange supported by protocols. Protocols are rules observed by all communicating parties. Their interpretation calls for intelligent devices, and this leads to the third mode of operations.

Personal computers and programmable terminals of all sorts have been in the background of recent developments in electronic mail. They will provide video (softcopy) supports, buffer the messages, interpret protocols, and generally interface between the line and the end user.

That is where (and why) electronic mail and videotex merge. Typically, the latter works in pages, promotes not only text and data but also graphics, and features color presentation.

The able use of these facilities calls for a significant amount of preparation, as we will see in the subsequent chapters of this book. A key feature in implementing electronic mail is that of learning how to use information. The data may be ready but not necessarily the executives. Yet, harnessing information and its understanding will be one of the basic premises of corporate management in the 1980s: Information is management's most important resource for productivity.

That is the reason why it is imperative that we concentrate on organizing and utilizing the information assets more effectively and efficiently. The corporations that will excel in the coming decade will be those that have the competitive edge in running the computers and communications systems in the way most useful to executives.

Neither the management information system nor office automation is yet a reality that we can purchase off the shelf. We have to use our imaginations in designing either one of them, and we must follow paths that will not limit us later. This requires a strategic plan showing where the organization is headed. The planning effort should:

1. Precede, not follow, equipment installation

2. Be based on an analysis of the functions performed in the various departments

3. Reflect the flow of information among departments and geographic locations

4. Account for the economic costs and benefits involved

Electronic mail is part and parcel of office automation. An effort in its direction should start in high-payoff areas with users who understand the concept and are trained in the application of the new technology to the office. Underestimating the planning time that must precede implementation is the main cause of difficulty in harnessing electronic power to our work.

It is no less true, however, that the prerequisite preparation for electronic mail is far simpler than that for videotex, which is discussed in the following chapter. Just as true is that electronic mail, in its various implementations, will be substituted for a good deal of the service currently provided by more classical means from voice telephone to the U.S. Postal Service.

Let us now look at the mechanics of electronic mail. They rest on the fact that computers, terminals, and communications facilities can effectively be used by businesses and private individuals to send and receive messages. The goal is speed of message delivery without interrupting the activities of the recipient of the message in the way the telephone does.

In this sense, electronic mail is not a totally new system; both facsimile and telex have existed for many years. Yet the latter two facilities have serious limitations as to their availability and serviceability. The limitations range from the slow pace and errors of the telex service to the lack of standardization of facsimile (though the CCITT Group I, II, III, and IV standards aim to correct that inconvenience).

The essence of these comments, however, is that the changing economics of computer technology and the increasingly widespread use of terminal equipment present new opportunities for electronic mail service. Not only does technology offer new challenges but the requirements for an efficient approach to document distribution are multiplying. One of the tough jobs in present-day business is the updating of organizational norms and the timely distribution of management orders. Electronic mail can do both jobs ably, and so can videotex.

Electronic mail has a potential impact on business communications methods that places it right at the center of our interest. There is, therefore, no wonder that companies are rushing in with electronic mail offerings. In the United States alone:

- On-Tyme and Augment, offered by Tymnet, are text-based.
- Telemail, by Telenet, also is text-based.
- IS1, by ISI, is a text message service like On-Tyme.
- Q-Fax, by RCA, is facsimile-based.
- Faxpak, by ITT, is facsimile-based.
- Faxgram, by Graphnet, is facsimile-based.

Text-based and facsimile approaches are technically quite different from one another. Facsimile, a nineteenth century invention, is a scanning process by which a page is copied and transmitted line by line. Speed and resolution depend on the group to which the sending and receiving devices belong. This makes it feasible to transmit annotations, pictures and diagrams. That is not true of the typical text-based unit, which transmits and receives encoded characters.

The six electronic mail offerings to which reference has been made are public, but there are also in-house systems through local area networks like Cluster/One and PLAN 4000 (by Nestar), Omninet (by Corvus), Ethernet (by Xerox), ARC (by Datapoint), Wangnet (by Wang), Videodata (by 3M/IS), and so on.

Typically, with public offerings the user can choose one of the several levels of service: Messages are held passively; rush messages are delivered by dial-out; other messages are delivered at preselected times. Priority of delivery is feasible. Multiple priority levels are desirable to many companies to avoid the unnecessary interruption of the recipient's work and make better use of the switching system.

At least three levels are useful: urgent, standard, and overnight. For standard messages, a delivery time of up to several hours will generally be appropriate, and overnight hours will permit spare transmission capacity to be used outside normal hours.

Such systems have in common the fact that they are software-based and are, therefore, computer-supported. In evaluating their performance, particularly for in-house solutions, the user would be wise to look at their software first (Fig. 4-1) and then examine:

- His own premises, the procedural structure on which the use of electronic mail will rest

- On the manufacturer's side, the hardware behind it

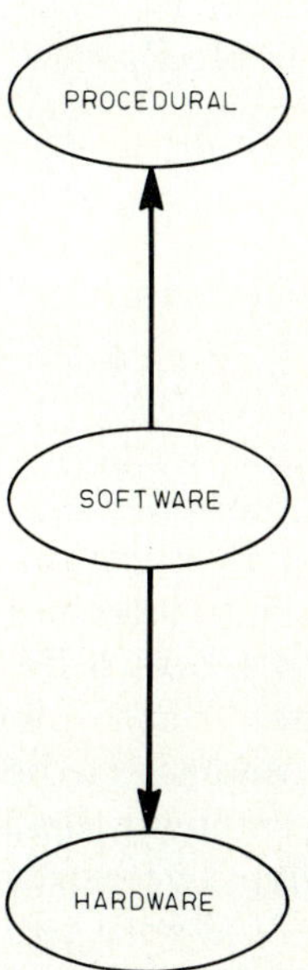

Figure 4-1 A properly done systems study should focus on software. Only after its specifications have been established should procedures be revamped and the hardware be chosen.

Since most companies already pay for fairly expensive computer gear and private telephone lines serving a classical realtime setup in data communications, several computer manufacturers have written the software necessary to help their clients implement electronic mail. In one of the companies I am working with, we have (with Honeywell mainframes) implemented the message-switching capabilities of the front-end software as a first approach to electronic mail. The following are the prerequisites at the systems level:

1. Automatic dial-out capabilities to be realized in collaboration with the telephone company

2. NPS map strings to control the automatic dial-out

3. The use of Autodial to avoid private lines.

As an alternative, the software can journal the messages and have the terminals call in. But that is a more restricted approach to the issue. Not that the basic software prerequisites are there. Not only does it support message switching but the now implemented OS release also has an electronic mail facility including mailbox file at the user file. This feature allows for specialized headers to be added to the mail and acts as an integral part of timesharing. The key word is MAIL.

Using Electronic Mail

The following is an example of the implementation of electronic mail, and the executive looking for the big picture may wish to skip this section. Its reason for being is to present a contrast with what we will soon be discussing about videotex. For that reason, it has purposely been kept simple, close to the fundamentals of the original offerings.

The MAIL subsystem permits a timesharing (TS) user to communicate with other users. The subsystem employs an appropriate interface to inform users of MAIL and uses another interface to access a file (MAIL.BOX) subordinate to the ability to list messages from other users, delete messages, or provide message information. Messages to be sent and ports to receive MAIL may be predefined on permanent files.

The MAIL subsystem command options consist of five key words which are interpreted as "action words" and are augmented with two specialized options [asterisk and equal sign (=)]. The action words are CREATE, DELETE, LIST, LISTL, and LISTD. Any subset of the key words string may be used to identify the command.

The total MAIL command syntax is made up of nine groups:

1. MAIL

2. MAIL*

3. MAIL CREATE (or any subset of CREATE)

4. MAIL DELETE (or any subset)

5. MAIL LIST (or any subset of LIST)

6. MAIL LISTL (no subset permitted)

7. MAIL LISTD (no subset permitted)

8. MAIL UID TEXT

9. MAIL =CAT/FILENAME

MAIL, followed immediately by a carriage return, implies that the user wishes all messages sent to be displayed at the terminal. If no messages exist, the statement NO MAIL AT THIS TIME will be displayed. Also, this syntax sets a flag which subsequently allows the message YOU HAVE MAIL to be sent to that userid (user identification) by TS LOGON. This functions as a "receive mail" feature and sets a flag so that the YOU HAVE MAIL message is displayed, if it is appropriate, during subsequent log-ons.

MAIL* is functionally similar to MAIL(cr) except that the NO MAIL AT THIS TIME is suppressed if there are no messages in the file. This syntax also resets the above-mentioned flag so that the YOU HAVE MAIL message is not sent to that userid during subsequent logons when there happens to be a message in the user's MAIL.BOX.

MAIL CREATE (or any subset of CREATE) directs the subsystem to create a MAIL.BOX file for the user. This file has an initial size of two links (640 words) and will be increased in size by the subsystem as needed for larger messages up to the maximum limit determined by the space available.

MAIL DELETE (or any subset) deletes all messages sent to the user's MAIL.BOX file.

MAIL LIST (or any subset of LIST) directs the subsystem to list the header information for each message in the user's MAIL.BOX file without displaying the actual message content. The header information consists of the total number of messages sent, which userids sent them, the number of characters in each message, and the date and time each message was sent. MAIL LIST UID displays the header data along with all messages sent to the mailbox from the userid specified, along with the header data.

MAIL LISTL (no subset permitted) causes the header data to be displayed to show how many messages have been sent to the requesting user. Also displayed are the date, time, and userid of the latest messages sent (those with the current date).

MAIL LISTD (no subset permitted) lists all header data of messages sent to the user and concurrently deletes the messages.

MAIL UID TEXT directs the MAIL subsystem to send the message text as entered to the userid specified. The user sending the mail will be informed if the receiving userid did in fact receive the message. The message initiator will receive the message USERID NOTIFIED, where USERID is the userid of the receiving user.

MAIL UID directs the subsystem to prompt the user for the message input. The user will be prompted until a null line is encountered. (The response is carriage return only.) Upon receiving the null line, the entered message will be sent to the specified userid after the sender has had an opportunity to verify the message's correctness.

MAIL =CAT/FILENAME prompts the user for input and sends the text to the userids specified in the file (FILENAME). The command MAIL =CAT/FILENAME TEXT sends the entered text to all the userids specified in the file (FILENAME).

When a file is to be used by the MAIL subsystem, it must be in standard system format and contain ASCII records. The use of the contents of the file is determined by whether the FILENAME is positionally the "first" or "second" option on the line. In item 9 the FILENAME specification is the first option and denotes the file contents as userids.

The maximum size message that can be MAILed is 4K (4096) ASCII characters, including control and end-of-line (end-of-record) characters, which are not displayed at a terminal.

Let us conclude with the statement that electronic mail is evolving. Intelligent terminals make it feasible to handle all of the protocol, for instance, by typing in a *phase number* like 2.1. Simple, easily available programs for personal computers can be called upon to add a carriage return at the end of an outgoing text line and a response to host prompts. But while electronic mail develops, the offering of graphic sets, color, and page handling by videotex is a major step forward.

The Role of Videotex

We said that videotex (or Viewdata) is the name of an interactive message service which uses domestic TV receivers, suitably modified or supplemented, as terminal equipment. Its implementation rests on four pillars:

- First, the telephone lines are similar to those of electronic mail.

- Second, the protocols are not the same.

- Third, instead of hardcopy being the main suport for data transcription and registration, input and output are video (though a hardcopy attachment is possible).

■ Fourth, files of messages exchanged (or to be exchanged) are kept in the VDB and are accessible to authorized people as the need arises.

As usually happens in new information technology fields, videotex appears to be different things to different people and is certainly not the same thing to all people. To some it is merely an interactive information retrieval service, and to them nice pictorial graphics are not essential. To others it is a new medium for reaching out to the consumers. In that sense, advertising, home banking, and video games are important elements.

Whatever our individual ideas of videotex may be, there is some general agreement that a videotex service permits the user to communicate with alphanumeric text and data as well as through graphics. In the years to come there will be abundant evidence that the strength of a videotex service rests on color and pictorial graphics.

Because videotex promotes an interactive exchange, this is the proper place to talk of rule 3 in the design of modern information systems. It is particularly applicable to communications-intensive activities (as every DIS should be) and has its origin in the Arpanet experience. "A message must be handled as a file in the database. Every file must be designed as a message."

Rule 3 should guide the design of all databases, and most particularly the VDB. It should also be a guiding principle in the implementation of interactive systems, with design choices able to assure the easiness of query, data entry, and other activities at the user's terminal.

One of the key attractions of videotex is that it is very user-friendly. This encourages the end user to try new features and to experiment with ways of getting something done more quickly or conveniently. That is particularly important to a person preparing documentation of any type, sending memos to more than one location, indexing files, abstracting documents, or simply exchanging mail. The same thing is true of answering management requirements for interactive reporting.

In a way similar to electronic mail, users are able to retrieve, via the common carrier telephone lines, other interactive networks, or in-house systems, data stored in computer-run databases and have it displayed on their TV receivers or video business terminals. They can access textual information by conducting systematic searches using simple numerical keypads. Individuals with slightly more complex terminals may compose their own information to be stored in the VDB for access by others.

In this sense, users are able to communicate with each other and interact by seeing the same images displayed on their individual TV screens. These capabilities, which bring advanced technologies to the home user, could have a profound effect on communications and on society. Such systems allow the introduction of many different services to home and business application. Electronic newspapers, electronic mail, electronic advertising, electronic games, and business transactions are supported by the new communications technology.

This business- and consumer-oriented experience started in March 1979 when the British Post Office (BPO) launched the Viewdata service, which employs an efficient combination of the telephone and the TV set to provide people with instant access to millions of words of information and entertainment. Stored on computer-supported VDBs, this information is called up over the phone and displayed in words and simple diagrams in videocolor.

Because in England the law implies that no company can take the name of the product it markets, the BPO public offering is known as Prestel. In France, Antiope and Télétel are the terms being used; the second is connected with the electronic phonebook experiment. Under the old standard, both have used alphamosaic protocols. Telidon, the Canadian offering, which is becoming the North American standard, is a bit-based process. We will return to these matters.

As a British assessment puts it, videotex is the world's first individualized mass medium, a publishing system more than an information retrieval exercise, an update medium (every second is a deadline), and a promotional medium (each page promotes subsequent pages) but not yet a fully interactive medium.

Invented at the Post Office Research Centre, Prestel has already given proof of the capabilities inherent in electronically handled data systems. Computers in different locations are being linked up and operated together in a network to provide Prestel with the capacity to handle more users than a single computer might and to do it reliably and in a reasonable time.

The first public service of its kind in the world was made available to residents of the London area. By September 1979, it was offered to businesses. The VDB is run by the BPO, but its content is property and responsibility of the information providers (IP). This is consistent with the policy the postal, telephone, and telegraph service (PTT) has followed with telephones. The common-carrier approach was taken because it would be politically difficult for the BPO to adopt an editorial stance toward information. That is not and has never been its business.

Other considerations also affected that decision. It would be administratively impossible to precensor all pages before publication (and would massively increase BPO costs). And, finally, the common-carrier policy would foster the sort of competitiveness which would directly influence Prestel's market success. Many users would not have come on board under a different arrangement.

The information contained in the pages of the database is prepared by IPs who, in several cases, obtain payment each time their information is accessed. Other information suppliers, such as advertisers, provide their information free of charge. Governmental bodies also supply free information as a service to the public. But the essential thing is that the IPs are not a government authority and are not regulated by laws other than those governing authors, publishers, and the daily press.

Sam Fedida, who is to be credited with this original development, most likely had a difficult time communicating to the eventual users the new avenues Viewdata opens to them. That is a recurrent problem in electronics at large (and in the computer field in particular), where ideas are ahead of the image the public has about the processes.

Accumulating Videotex Experience

The problem with most office and consumer products is that they do not provide what the user needs and wants. Fedida's design seems to have been made with the user in mind. With computers, everyone says it, but no one is paying attention to it. An easy example of this can be taken from the office: Whereas we should discourage the user from typing, word processors are doing exactly the opposite—they are designed around the keyboard. But, for messages consisting of framed choices, videotex can be operated through a keypad like that of a touch-tone phone.

Another basic consideration is the installation procedure. Not even minicomputers and word processors get installed by themselves. Someone has to install them, and that person should have DP abilities. But the public offering of videotex is reaching homes and businesses through the TV set and the phone line, and videotex calls for no more skill than turning on the TV.

The experience so far is eloquent. Videotex is a significant advance in an interactive communications technology, and the work of setting up the system proceeded at a surprisingly fast pace. In about a year and a half of concentrated work, the BPO experts:

- Chose the hardware characteristics and commissioned the system for production

- Developed the software for one of the world's biggest computer-based information retrieval projects

- Installed a fully tested operating system

- Developed, with the television manufacturing industry, more than a dozen types of Videotex-compatible TV sets

- Were the first to solve the technical problems of connecting TV sets directly to the phone network
- Assembled, with the information industry, a brand new approach to database management
- Overcame a host of novel problems in the fields of legal liability, editorial control, consumer protection, copyright, and privacy
- Put on the system first for a controlled group of users, then for the general public, and finally for business and industry

Prestel requires the user to be active. Until the user makes a decision, the same page will sit there. Prestel takes no initiative. Therefore, a key skill in Prestel design has been to encourage active response and thus augment the desire to view more pages.

The public utility provides general access to very large databases that are stored in central computers and organized as display pages. The system makes use of the home TV receiver as a still-picture display medium. The great departure from the regular television service is that the viewer has much more control over what information is presented on the television screen and has the ability to become not only a receiver but also a sender of information by creating text and images for insertion into the database.

The ease of communication is vital, because the computer industry's ability to produce more sophisticated machines was always well in advance of the user's ability to understand the products. In this sense, there is a reversal in trend, because videotex is so simple and low in cost that it can give its user—whether business, industry or financial institution—a leading position in information systems. And such position will be of vital importance in the 1980s.

Critical to a public videotex network is the role of the information provider. In England the information is provided by organizations such as British Airways, British Rail, Central Statistical Office, Consumers Association, Thomas Cook, Fintel (the *Financial Times* and *Extel*), *Good Food Guide, Investors Chronicle*, Meteorological Office, W. H. Smith, and the stock exchange. And there is a great demand for more frames on the VDB from both exisiting and new IPs.

Such are the business opportunity projections made by some of the information providers that proceed with fairly sizable investments. For example, Sealink, the cross-Channel ferry operator, spent hundreds of thousands of pounds to equip 2500 travel agents with Prestel sets. One asset users have so far obtained is experience in videotex technology. The experience proves most helpful in the developing application of a public facility and in-house interactive information systems.

The user, whether business, industrial, financial, or private, receives realtime answers to queries and can also transmit information to central databases or other users. Individual ideas like videotext and teletext will eventually be joined by other developments which will be built into the television set of the future.

Another major development is computer-assisted education. In its first period it will be aimed not at replacing existing methods of training, but at adding to them. With options open for both in-house systems and a public utility, schools and businesses can build their own training courses and run them in three ways:

- On stand-alone machines
- Through an internal private videotex system
- Through the national public utility service

Three features stand out in the development of the Prestel computer network:

- It gives users local call access to the information bank through a ring of local centers, thereby cutting out the expense of long-distance calls.
- It provides a dynamic base for the information providers, who can augment and update their individual text and databases quickly and easily without having to call up each local computer in turn.
- It ensures that once information providers have updated their databases the new information is duplicated on all the local access centers.

Immediacy of updating, but also simplicity and ease of use, are Prestel's outstanding features. The casual reader of this book may think: "And what's in it for me? Prestel is operating in England. What's the fallout over here?" To answer that fairly legitimate question, all we have to do is to recall that videotex is offered both as a public utility, as is the case with Prestel, and as a private, nearby, in-house system.

AT&T seems to be particularly interested in the process. Since May 1981 AT&T has revealed its videotex design. The announcement did arouse resistance by newspaper publishers, who do not want the telephone giant to get into the electronic information business at all. They fear that AT&T will steal their classified advertising by offering on the videoscreen the kind of commercial information not available in the Yellow Pages directories. Electronic directories could be updated daily, offer news of prices and bargains and, unlike the Yellow Pages, compete directly with newspaper classified advertisements.

Many other organizations in the United States have announced plans for a public service (GTE, Knight-Ridder, Times Mirror, CATV outfits), but public offerings are only one part of the market. Nothing forbids a bank or merchandising or industrial firm from setting up its own videotex system, whether for internal use or as a company-operated and -maintained network supporting metropolitan or regional subscribers.

Cooperative efforts also have their place. IBM, CBS, and Sears, Roebuck are throwing their weight behind videotex services in a new joint venture. The three firms intend to offer a variety of products: shows from CBS, merchandise from Sears, brokerage from Dean Witter Reynolds (a Sears subsidiary), home banking from financial institutions, and, most likely, telesoftware from IBM.

But whether a firm acts as an information provider to a larger, public network or runs its own system, the preparatory work is a *must*. As we will see in the appropriate chapters, this includes the structure of the menus, the routing pages, and the infopages, as well as decisions on the latters' content and framing. Above all, the decision to use videotex calls for a new image in message handling, including management's conviction about its advantages.

Videotex should really be seen as the color-and-graphics development of electronic mail, able to reach the ultimate consumer in his living room. Videotex and electronic mail have several characteristics in common:

1. They have high utility.
2. They are a good way to introduce people to new technologies from office automation to home banking and telecomputing.
3. They are easy to use.
4. They are fairly inexpensive.
5. They require only simple training.

Above all, videotex and electronic mail are a common communications vehicle crossing all borders.

A Worldwide
Videotex Standard?

*"The turning point in our user relationship
came when we got so many of our own people
into the user community."*
VANDE WOUDE, *Computer Decisions*

Videotex is a competitive environment; and when we talk of a competitive environment, a distinction must be made between in-house offerings and public media. The former require greater experience, about which we have been talking throughout the book; the latter usually offer packaged solutions and have on hand some information providers which for a fee can help newcomers get started. Still, the most fundamental aspect is standards—and the lack of them.

To start with, there was no international standard until February 1983. But, as with computers and data processing gear in general, de facto standards evolved with market acceptance. Here, it is not IBM which makes the rule. In Europe the "alphamosaic standard" became established because the various telephone companies (PTTs) adopted the British Prestel as their basic software, albeit with certain modifications.

That is the story of the Dutch, German, Swiss, Austrian, and Italian PTTs. The French and the Swedish PTTs have a different system. However, in terms of graphics (more precisely, semigraphics) it is not that much different from Prestel, except that the French system needs 2 bytes of storage for every byte on the screen to avoid Prestel's "black holes." With the U.S. telephone companies (AT&T, GTE, and so on) adopting the Canadian Telidon standard, the balance swung that way. The big market for the videotex services is the United States.

Yet, although the technical details may vary, all videotex developments point in a certain direction: that of the evolution of a new type of communications services. Table 5-1 dramatizes the evolutionary process in the background and traces the steps we experienced from the written to the visual word and as a function of the writing and reading surface. This trend, with the notable tendency toward the integration of voice, text, data, and image in one communications network, is in full swing.

The turning point in the adoption of a graphics standard has been the development by Canadian Bell of a more sophisticated approach with curving line (than Prestel's alphamosaics). Telidon uses a bit-oriented protocol which is evidently more precise. For the end user it also calls for a more expensive terminal, though the domestication of the microcomputer can change that.

The user-friendly approach is another key point. In the current race to persuade customers to overcome their resistance to the many kinds of communications technology, it is most likely that the winners will be those whose machines do not bristle with difficulty but do appeal to the average person as easy to use.

Table 5-1   The Evolution of Means of Communications

	Written word	Typed text	Telex	Micro-fiche	Facsimile tele-copier	Video-phone	Video disk and video tape	Classic real-time	Video-tex
Writing surface	Paper and pencil	Alpha-numeric key-board	Alpha-numeric key-board	Film	Paper	Camera	Camera	Alpha-numeric key-board	Numeric keys only
Trans-mitter's storage	Paper and pencil	Paper	Paper tape	Film	Paper	CRT	Plastic	LSI	CRT
Data carrier	Paper and pencil	Paper	Radio or tele-phone	Film	Radio or tele-phone	Radio or tele-phone	Plastic	Radio or tele-phone	Radio or tele-phone
Receiver's storage	Paper and pencil	Paper	Paper tape	Film	Paper	CRT	Plastic	LSI	LSI
Reading surface	Paper and pencil	Paper	Paper	Projec-tor	Paper	CRT	CRT	Hard-copy	CRT

In-House Offerings and Public Media

Videotex offerings can be either private or public (Fig. 5-1). Public videotex can be supported by the telephone utility (AT&T, PTT), by a service bureau offering, or by a value-added network (Telenet, Tymnet, and the like). A private videotex system can run on a local **area network** and personal computers or a dedicated mini or share **mainframe** resources. The efficiency of the latter is limited. It can support a relatively small number of users, and its response time is slow because other applications compete for it.

It is well known by experiment that mainframes are not well suited to supporting a large number of terminals. (At least one major bank has tried running mainframe videotex and abandoned it.) The mainframe is neither an elegant solution nor an efficient one. Long response times destroy the reason for using videotex and disgust the end user.

One must, however, be careful in selecting the in-house system. The logical mechanisms supported by vendors are different and often incompatible with one another, and there are different releases within the same system. In other words, the facilities are not necessarily universal: Prestel has one type of frame (page), whether data or routing. The Philips software supports two types of pages, one for routing the other for data.

Other differences are in communications. The original Prestel soft-

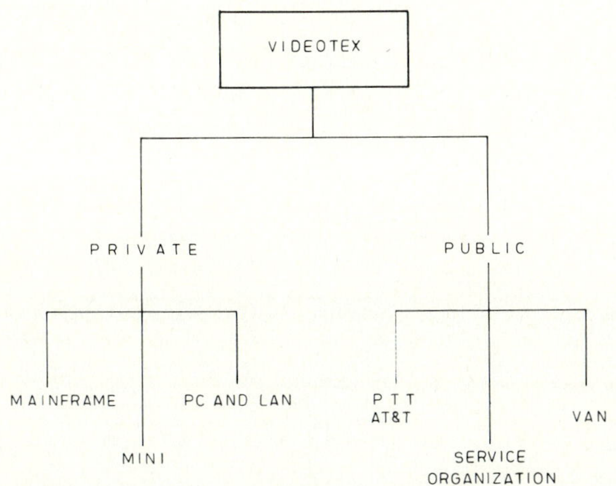

Figure 5-1 Videotex services can be divided into two main classes, private and public, each of which is further subdivided.

ware worked half duplex and used the return wire to send data. That made the transmission from the public utility computer at 1.2 kbps, while the response frame (from the end user to the computer at the switching center) traveled at 75 baud—the telex (TWX) speed. American releases of Videotex use full duplex lines and can work both ways at 1.2 and 2.4 kbps.

When we use nonintelligent terminals, the hookup of the TV set to the telephone line changes nothing in that it continues to operate on and off as a TV. And it can receive at as high a transfer rate as broadcast stations (or satellites) can transmit. That is important because the principles of videotex and teletext are quite similar, as Fig. 5-2 shows.

The interactive features of a videotex system can be nicely complemented through the broadband capabilities of teletext—sending written information for viewing on a videoscreen through microwaves or CATV. The early teletext systems, like closed captioning for the deaf, started by inserting digital data into the unseen scan lines at the top of the TV sweep. As the applications expanded, the fact that those unused lines can hold only a few pages of text per second became a bottleneck.

Expanding into a full cable channel allows the data stream to be larger than 500 pages per second, and the selected number of pages in the stream depends on the acceptable waiting time between request for and display of data. Among current offerings, Zenith's Z-TEXT cable service offers 40 text services with an average of 125 pages each. (Zenith

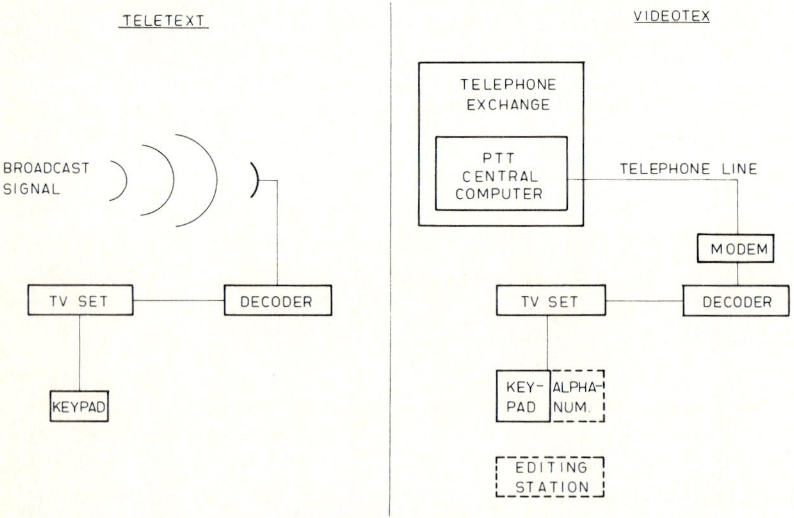

Figure 5-2 On comparing teletext with videotex, we see that there are significant similarities at the receiving end.

will supply only the hardware, and it is encouraging cable operators to offer such text services as news, professional information, business reports, cable-TV guides, classified advertising, and shopping information.)

To expand a CATV service significantly without an addressable hub requires attention to three technologies: picture scrambling, addressability, and two-way communications. And here is the prospect of the *wired city*. By the mid-1980s, it is estimated that there will be 70 million homes in America worth cabling. The cable (whether coaxial, optical fiber, or waveguide) carries clear signals which can be a TV show or an *informational system:* teletext, videotex, electronic mail, newstext, and other communications. Already TV stations transmit (invisibly and during their regular broadcasts) 200 pages of textual data—airline schedules, theater listings, news headlines—to viewers' TV sets.

Currently, a special control lets a viewer select any desired page of data and freeze it on screen (CBS, with its station KMOX–TV in St. Louis, is experimenting with such a system). With earth station receivers dropping sharply in price and coming within range of the average cable operator, the once exotic and expensive devices only governments or huge companies could afford might easily become commonplace.

It is no wonder that so many well-entrenched companies and individualistic entrepreneurs are scrambling to get in on the ground floor of a technology that will change the face of the TV industry in this decade. "It's as if the entire industry has come off the highway it has always known and onto a traffic circle with six or seven different exits," said a telecommunications executive. "A few companies are circling, trying to decide which road to take to get farther ahead; others are driving down new roads—but they are watching in the rearview mirror to see where the rest are heading, and they are wondering if they've done the right thing."

Information-based companies (McGraw-Hill, *The New York Times,* Knight-Ridder, Times Mirror, and Time Inc.) are looking into the acceptability of personal computers that can use a TV set as a display screen to sell access to their databases to the business and home markets. New services like *narrowcasting* will serve small, specialized audiences.

The telephone companies may be fighting back with new products. Advances in robotics, coupled with developments in artificial limbs, might lead to an entirely new type of communication device. References point to the sudden growth in the industrial and home robot market as evidence of possible future interest in telephony. (Manufacturing plants are turning to robots to perform a number of tasks including machining, welding, painting, and other "undesirable" jobs that have been performed by humans, while computer companies slowly enter into the robot market.)

Research in the nuclear industry, where manipulator hands for nuclear weapons and nuclear power plants are in widespread use, can be related to units such as Feel-a-Phone. In some nuclear facilities, very complicated processing tasks are performed on highly radioactive materials. The operator works the machines by remote control through robot manipulators which are guided by the operator's finger movement on control rings.

Do we really need such devices? Let us turn the clock back 100 years and ask ourselves if there is a need for such "exotics" as photo cameras and hand-operated telephone boxes. The answer then was *no!* Today, the answer to the same question is *yes!*

Alphamosaics and Alphageometrics

In view of the coming widespread use of videotex services, standardization is a legitimate concern. Are we going to have a single standard that makes it feasible to communicate over the lines with another user no matter the kind of equipment employed or the network to which the equipment is connected, or will we be confronted with a myriad of protocols and character sets that make interconnection dubious and discourage wider use of the service?

To answer that question in a valid way, let us first look at the fundamentals. Videotex systems employ pages of graphics and text, and each page corresponds to a display image. As the system developed, an array of display techniques began to be used: alphamosaics, alphageometrics, alphaphotographics, and dynamically redefined character sets (DRCS).

Alphamosaics are based on a gridlike division of the screen, typically into 40 columns by 24 rows. Each of the cells can contain any one of a fixed number of predefined graphics characters. *Alphageometrics* correspond to vector generation displays: Images are composed of points, lines, arcs, polygons, and text. *Alphaphotographics* are implemented through raster graphics. In this case, codes for *pixels* (picture elements) are transmitted sequentially. At the level of the presentation level protocol (PLP), another important definition is the *pel,* the logical presentation element. *Dynamically redefined character sets* allow each page to use a tailored character set by transmitting first the cell display corresponding to each code and then one or more grids of codes that define complete display images.

The original viewdata (Prestel) offering worked in alphamosaics (semigraphics) by using a byte-level protocol. The competitive Canadian

Table 5-2 Comparison of Videotex Alternatives

Prestel (English)	Telidon (Canadian)	Télétel (French)
Main difference is the graphics. Textual data seems the same.		No Prestel-like "black holes" for changing colors.
Prestel: alphamosaic	Telidon supports alphageometric, allows drawing from coordinates, and uses encoding device like a scanner Resolution is a function of the receiving unit	Different character standard than Prestel (Both are CCITT supported.) Two bytes per character

With the Japanese solution it is possible to pick up a computer-made picture.

Telidon supported bit-level handling and hence alphageometric presentation. A brief comparison between Prestel, Telidon, and Télétel is made in Table 5-2.

Figure 5-3 gives a quick look at the graphic facilities obtained by use of the *alphamosaic* and *alphageometric* techniques. Typically, in an alphamosaic setting the character is divided into six spaces, though in a version made by the British for GTE the character was divided into eight parts. Specialized computer graphics terminals have always been alphageometric. Precision is much greater, and so is the cost.

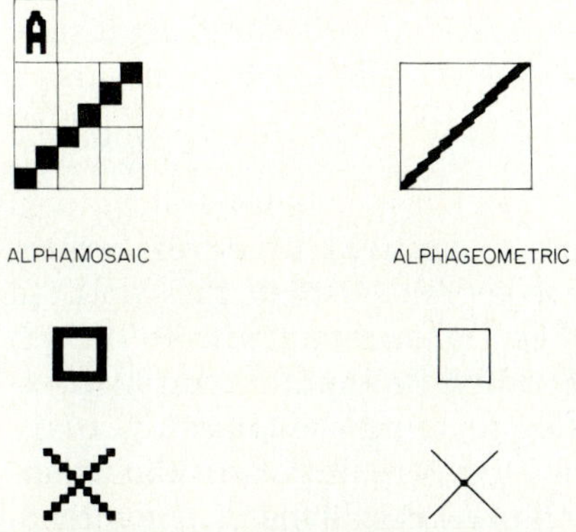

Figure 5-3 The alphamosaic and alphageometric presentations of basic graphs are the solutions presented by Prestel (semigraphics) and Telidon, respectively.

Again, in the original videotex offering (prior to the establishment of CEPT-PLP), the growth of character-oriented systems was restricted not so much by their display methods as by their image description techniques. Prestel and its clones transmitted images as sequential pieces of a picture consisting of 24 lines and 40 columns based on the European 625-line TV formats. Graphic images were constructed from specially identified coded graphic characters fitted together as individual pieces of the complete picture as in a mosaic. By extension, terminals employing that image description and display technique have been referred to as "alphamosaic."

Once the alphamosaic method is accepted, it is impossible to send higher-resolution images without knowing the display capabilities of the receiving terminal. The problem caused by lack of a standard is that databases may have to store multiple versions of each image to suit the proliferation of terminal configurations that may occur as technology advances and better-quality graphics are demanded.

For greater flexibility, the Canadians have insisted, images must be described at the VDB in such a way that they are completely independent of:

- The data access arrangements at the central computer

- The characteristics of the communications medium

- The display terminal construction and resolution capabilities

One solution to this problem has been the handling of graphic images containing geometric shapes, including a means of describing photograph-like images. Textual messages could be transmitted by means of standard procedures using 7-bit ASCII characters. Thus:

- Terminals capable of receiving and displaying structured geometric shapes and text might be described as *alphageometric* terminals.

- Terminals capable of displaying photographic-like images would be called *alphaphotographic* terminals.

Terminals of both designs should include circuitry able to interpret picture description instruction (PDI) codes and display the captured image. Furthermore, a terminal might be alphageometric even though the display has an alphamosaic format when its input circuitry interprets geometric image descriptors before forming the output display.

More sophisticated terminals contain a high-resolution display memory with a location associated with each pixel displayed on the screen. Such terminals are capable of presenting photograph-like images as well as alphanumeric characters and graphic images, and the graphic images could be displayed with much finer resolution. [However, Prestel supporters answer this argument by saying that even with character (byte) solutions you can do more things with different levels of resolution and grade scaling.]

Technically speaking, the method proposed by the Canadians for storing and communicating visual images and textual information was through the use of PDIs. For reasons of future flexibility it was suggested that the resolution of stored images is virtually independent of terminal configurations, communication networks, and database construction.

The Canadian method is to describe the contents of composed images with graphic descriptors of each object in the image. Each object, and its position in the image, is usually presented with accuracy, and so the display resolution of the final image should be suitable for most applications in the foreseeable future. This resolution is about 10 times what is required for present-day television receivers, but the high-resolution word processing equipment now available can resolve images approaching half the defined resolution.

Even higher resolutions could be defined without radically affecting present-day equipment. The extra bits required to define images with such high resolutions fit naturally into the construction of the PDI set and can be used to handle images for facsimile or form processing equipment. In other words, the PDI set is not necessarily restricted to videotex, but new terminal equipment which could interpret the PDI codes would be required. (Essentially, pictures are described with a set of geometric drawing primitives, such as line, polygon, and arc, specified in various positions in the total picture to be displayed. Similarly, text is presented as a group of characters to be displayed at certain positions on the screen. In that way, the PDIs can describe practically all textual and graphic images. Images or parts of images which cannot be described with geometric drawing primitives can be specified in the photographic mode of operation.)

Only seven basic instructions, each followed by high resolution, are needed to describe practically all graphic images. Four of the instructions reflect objects of structured image in a geometric mode: lines, circular arcs, rectangular areas, or polygons. These basic geometric primitives were chosen because of their simplicity.

A fifth instruction indicates that the data following is in bit form, that is, the photographic mode, for images the structure of which cannot be defined even by numerous logical drawing primitives. A sixth instruction defines the position of an object. It may merely indicate a movement of the display beam prior to transmitting one of the primitive commands, or it may be used to plot points in random fashion on the display. The difference is indicated with a specific bit in the command word structure. The seventh instruction is used for control; generally it sets a status register prior to sending other instructions. For example, one control command may be to set the value or color of an object.

A command is also required to change from graphic to alphanumeric mode for textual messages. This function is performed by the shift-in (SI) instruction from a set of control codes universally used in the transmission of textual messages. Similarly, the shift-out (SO) control code is used to change back to the graphic mode. The alphanumeric mode is the mode of operation entered by default, or the way in which a terminal should operate when first switched on. In that way, a subset of the PDI code can be used for business or simple alphanumeric terminals which may respond only to textual information. (The PDI instructions, associated data words, and alphanumeric characters are contained in 8-bit bytes. One bit of each byte is used for parity to comply with present standards of transmission over conventional data networks.)

We conclude this discussion of the factors behind Telidon by stating that, in a general sense, the user's terminal mode is based on a separation of the visual content and structure of the infopage. The latter is layered, and each of the three layers is independent of the others. The three are:

1. A full-screen background layer which may be partitioned into rows and columns

2. A defined display area background layer

3. A defined display area foreground layer

Though the logical candidate for the alphageometric and alphaphotographic modes is layer 1, for a specific application it is appropriate to identify the layer to which the modes belong. With that background we can now look into the international standardization effort.

The CEPT-PLP Standards Effort

The fact that there are three different modes of graphics presentation—alphamosaic, alphageometric, alphaphotographic—does not mean that a uniform mode could not be based on one of them or, better, include the other two as options. Indeed, international videotex standardization efforts have concentrated on the major coding schemes, with a trend toward merging them into what is now broadly classified as the presentation level protocol (PLP). This work has resulted in Recommendations F 300 and S 100.

Such efforts have seen the emergence of two regional videotex standards: European and North American. The latter has been jointly developed by technical groups of the American National Standards Institute (ANSI) and the Canadian Standards Association (CSA). The standards are both alphageometric and alphamosaic. It is possible that other regional standards may be defined to include alphaphotographic capabilities.

The European-North American agreement, signed in Geneva in February 1983, provides an exchange mechanism between the eight graphics sets of the North American PLP standard and the seven graphics sets of the European CEPT, and the scenarios which led to this choice are interesting. Four main alternatives were considered:

1. A single worldwide international standard for videotex

2. No international standard

3. Regional standards as subsets of an international accord

4. A single international gateway standard

An analysis of those scenarios indicated that (1) is totally impracticable and should be discarded, (2) involves no international standardization and imposes a very heavy burden on conversion and should, therefore, also be discarded. Alternative 3 was considered realistic and the best matching with the current environment. Alternative 4 could be simplified to (3), or else a monumental task remains in defining a neutral abstract international standard.

One of the reasons the first alternative was found to be impracticable is that it would require all terminals to implement all languages and all modes of coding pictorial graphics. That would be the price for the establishment of a single worldwide standard which is a superset of all national or regional standards and is implemented by all terminals and VDBs. If when the idea of an international standard came up, no country or region had had any standard at all or had not made an investment in any system, then it might have been feasible to start from scratch and develop a single worldwide standard independent of any national or regional system. But by 1981 that approach was already impossible. Commercial or advanced field trial systems were in progress in many countries, and large investments had already been made.

The chaotic situation and full incompatibility represented by alternative 2, no international standard, was enough to discourage inaction. Were that approach to be accepted, videotex terminals in a given country would be able to communicate with and retrieve information only from databases within that country. Videotex interchange between any two countries would have to be via a complex network of gateways with a very heavy conversion burden.

Alternative 3 is to have regional standards as subsets of an international standard. What this means is that every videotex terminal in every country in a region should be able to communicate with and retrieve information from every database in every country in the same region. Videotex interchange between countries in any two regions would have to be through gateway conversion. For this alternative to be practicable, the condition is that all countries within one region will use the same basic graphics sets and the same pictorial techniques (mosaics, geometrics, photographic, or any combination thereof). If that is so, we no longer talk of a single worldwide standard. Instead, we talk of regional standards as subsets of an international standard. This is very much the structure of Recommendations F 300 and S 100.

By alternative 4, a single international gateway standard, every videotex terminal in every country might have been able to communicate with and retrieve information directly from every VDB through a protocol (international standard) to be used for gateway conversion. This virtual gateway would, of necessity, contain a superset of all the functions and specifications of the current national or regional standards.

Alternatives 3 and 4 have a common ground in that each calls for a superset of current national and regional standards. If this superset approach is not followed and an attempt is made to develop a neutral abstract international standard based on tearing apart one or more current national or regional standards, then international agreement would be very difficult not only to achieve but also to maintain after it had been achieved.

It was along that line of reasoning that CCITT produced recommendations F 300 and S 100. They contain two "options," one alphamosaic and the other alphageometric. The alphamosaic option contains two modes, one serial and one parallel. All three coding schemes have in common Latin-based alphabet coding, but even pure alphanumeric textual information cannot be freely interchanged among the three schemes. A third option, the alphaphotographic, also is identified in the recommendation.

The following appear to be the implementation realities. A unification of the serial and parallel alphamosaic modes, with certain enhancements such as the addition of smooth mosaics, is taking place in Europe. In Canada and the United States a common draft standard, Videotex/Teletext Presentation Level Protocol Syntax (North American PLPS) has been developed by technical groups of the national standardization bodies of the two countries.

NA/PLPS is built upon the alphageometric option of S 100. It contains a mosaic shape table which is a union of the serial mosaic and parallel mosaic tables of S 100. It also contains:

- Color mapping
- Controllable drawing line width
- Continuous character size scaling
- Programmable texture masks: unprotected fields, partial screen scrolling, incremental picture description instructions, macros, and DRCS

In Japan, elements for the alphaphotographic option are being developed and some contributions have already been submitted to CCITT. Concurrently, advanced and large-scale field trials, as well as provision of commerical services, are taking place in Europe, North America, and Japan. A great deal of investment is being committed. Hence, changes in standards affecting services can be quite costly.

Let us now look into the purpose of the PLP recommendation which describes the PLP for use between videotex terminals and VDBs. (A separate recommendation will describe the protocol to be used when interworking between VDBs.) The scope is to describe the display functions, define the character and control repertory, identify the coding, and elaborate the management functions which may be used to enhance the display.

The recommendation puts in perspective the need to define the management data protocols as well as the coding structure for distinguishing between data used for different types of displays and management functions. The resulting recommendation also had to be as nearly compatible as possible with existing recommendations and standards:

- CCITT Recommendation V3
- CCITT Recommendation F 300
- CCITT Recommendation S 100
- ISO Standard 2022 (Rev 79)
- ISO Standard 6937
- ISO Standard 6429.2

The reason for observing those recommendations and standards is compatibility with existing structures. For example, several coded character sets in use today, such as the International Reference Version (IRV) in CCITT recommendation V3 (same as ISO 646-1973), will continue to be prime factors in the evolution of the videotex industry.

Thus, the goal of which we are talking includes functional compatibility of existing and future videotex systems, internetworking between videotex systems with unique national requirements, the provision of cost-effective services, the utilization of consumer television receivers and general display terminals for videotex, and a reasonable ease of interworking between videotex systems and similar services—tetelext, electronic mail, and so on. Equally important is compatibility with the open system interconnection (OSI) architecture.

Techniques have been developed for integrating videotex syntaxes of CEPT-PLP and others into an international approach that meets basic coding requirements. The idea is to use a presentation level protocol to switch between the different data syntaxes, such as International Alphabet Number 5 (the IRV), North American PLPS, and CEPT videotex. for that purpose, the syntax of the presentation protocol data unit (PPDU) is designed to be compatible with both ISO 2022 Code Extension standard and ISO DIS 7498 Open Systems Interconnection standard. The presentation protocol data units are made up of two parts:

- Presentation protocol control information (PPCI)

- Presentation service data units (PSDU)

Each PPCI acts as a header and identifies the type of PSDU that follows in the PPDU. Two types of PSDU's have been identified so far: control and data.

Syntax will be selected by using a presentation level management function invoked by means of a presentation protocol control information portion of the presentation protocol data unit. A time sequence of possible exchange of protocols and syntaxes can be established. PPCI is a generalized announcement of the protocol syntax to be used, and it would comprise an escape sequence of any agreed form.

Figure 5-4 illustrates the contents of the protocol data units and the service data units in the ISO Open Systems Interconnection Reference Model. The presentation service data unit (PSDU) will contain either of the following:

- The current presentation protocol syntax

- The available presentation protocol syntax for negotiation

The presentation protocol control information will announce which of the two types of PSDU follows. When a change in context is requested, the receiver can accept or reject the request. This conforms to the philosophy and architectural structure of ISO 2022 and of OSI. PPCI information is needed only to envelope the PLPS. The PPCI does not work at the level of the particular code tables, because that would create an alternative invocation structure conflicting with ISO 2022.

In accordance with that protocol, each presentation service data unit contains a particular type of syntax. Four types have been identified:

- National Data Syntaxes, such as ANSI X3.4 (ASCII)
- International Reference Version (IRV)
- North American PLPS vedeotex
- CEPT videotex

PPDU is so structured that other standardized and private data syntaxes can be added. Each data syntax may be used in sequence, although each is separated into different PPDU during transmission.

Eight sets of presentation level commands are necessary to support an equal number of basic functions. The first selects a particular presentation process from the above list of four choices. The second switches between 7- and 8-bit character sets. The third changes from one presentation device to another, as from video display to printer. The other five have the following functions: Suspend or terminate the current presentation process, select data transparency mode, negotiate terminal options implemented, select data encryption (SDE), and reset the entire presentation layer. Any of the eight commands is optional for terminal implementation.

As an example of the functioning of the commands, select data en-

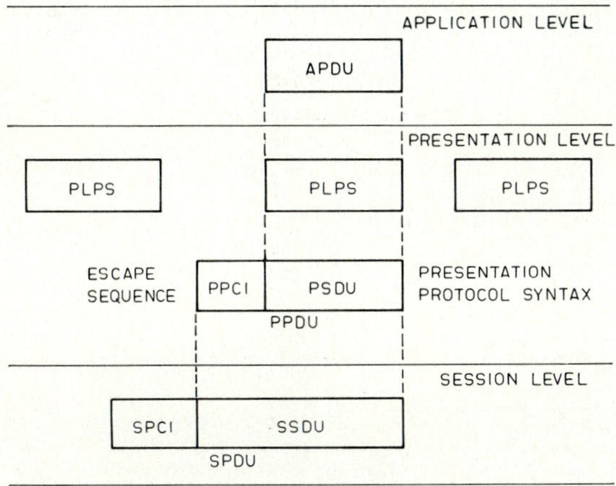

Figure 5-4 Working within the ISO/OSI layered architecture to accommodate presentation protocol control information (PPCI) and presentation service data units (PSDU).

cryption permits the sending process to indicate whether data in PSDUs is encrypted according to the parameter value. Either the host or the terminal can originate the SDE command. Its employment usually requires data transparency, as by using the PSDU length in the PPCI.

Classes of Terminals

The characteristics of the terminals which will be used within an interactive system are, in themselves, fundamental. The intelligent terminals to be employed should not only support the PLP which we discussed but also reflect at least three requirements whose needs may not be answered through exactly the same characteristics:

1. Classical data entry

2. Graphic editing

3. Reporting capability

The object of a terminal is to ease communication between man and information. A word processor and a copier can be a terminal, but an intelligent terminal will include a micro or will feed into a mini (Figure 5-5). Terminals may have microfiles (their own databases) and use telephones to communicate with one another and other computer resources. But in order to start a network there must be a terminal.

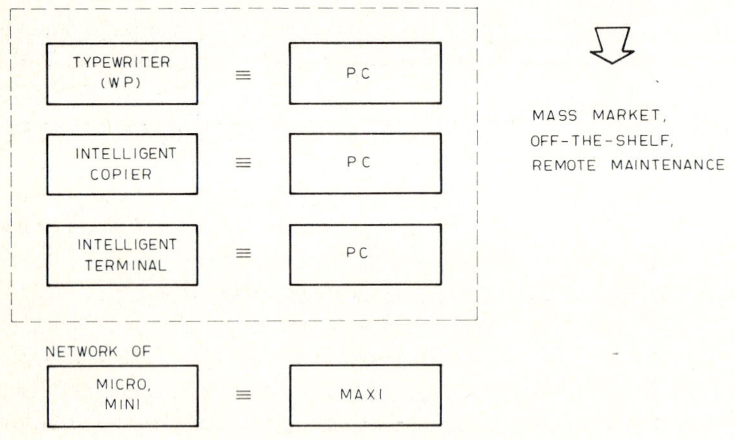

Figure 5-5 The evolution toward microcomputer-based terminals will see the integration of word processors, calculators, intelligent copiers, and other devices with the same unit: a personal computer.

Data entry characteristics must be considered, particularly those related to online data collection. Direct online text and data collection for office and industrial activities has been the goal of automation for years. Since videotex supports both the receipt and expediting of messages, there is no reason why we should not avail ourselves of this computer-supported activity in a companywide sense. Let us not forget that a key issue is the interactivity of the user with the information held in the system.

The text and data entry terminal can also help by way of graphics if there is a package in the background to convert tabular data to graphs. When we talk of graphics through terminals, we mean self-standing units able to handle graphs, text, and data through a full-color facility and permit editing of the graphical presentation before producing it.

Different types of charts should be available:

- Horizontal bar charts
- Vertical bar charts
- Line charts
- Combined bar and line charts
- Pie charts
- Word charts or textual charts

Also available should be colors to help emphasize and/or distinguish elements of a chart and a number of character fonts and sizes which can be used for any text or numeric labeling. Labels can be placed anywhere on the chart in either the vertical or horizontal position. For bar and line charts, there must be horizontal- and vertical-axis titles. They should be automatically positioned but capable of being moved. The system must provide automatic scaling of the axes to fit the data of the chart, a feature which may be overridden by the user.

Charts and data files should be easily handled through the WS. The chart definition can be stored with the chart data and labels, or the data and label can be stored separately. Different data sets and labels can be recalled and used with a chart, or conversely, the data sets and labels from a chart can be recalled and used with other charts. The workstation should have an integrated modem for communications; it should have at least a 1.2-kbps capability.

Software for query and reporting should be designed to answer effectively the management requirements of an interactive environment. That is the primary issue whenever we refer to videotex. Terminals in this category should be extremely user-friendly, fast, reliable, and forgiving. Variations are possible depending on the facilities supported.

Simpler (and less costly) terminals are those of telephone company design for electronic phone books. They can also serve as answer pages when implemented by cursor positioning to guide the user. Present-day improvements in cursor positioning include the use of mouse and joystick. A further-out possibility is actuation by voice. For office use, the editing facilities must be comprehensive, including the ability to:

- Move the cursor anywhere on the frame
- Delete or insert characters
- Delete or insert complete lines
- Enter graphics
- Assure graphics creation
- Move a block of text from one part of the frame to another
- Provide tabulation marks for column editing
- Do vertical tabulation for control character insertion
- Copy one part of a frame onto another frame
- View control characters on the display

The alphanumeric or strictly numeric keyboard input is a relatively easier requirement, but input units may get more complex as the details of the equipment shape up in response to market reactions. This is a double-edged sword, because product reactions are based on the acceptance of the service and the acceptance of the service is a function of the features of the first offerings.

Depending on use requirements, terminals must also have logical capabilities for a number of functions at local level; among them is bulk update. A keyboard operator typically creates 5 to 10 frames/h (frames per hour), whereas a microcomputer can create up to twice that number per second. For updating, a prime requirement of an online system, source data can be held in an external memory and the VDB created from it. Furthermore, as videotex use spreads, we may see user organizations relying on more than one videotex system. Each system may require a variation of text and database to take account of differences in code or VDB structure. That too calls for local intelligence in order to do an able job.

Although terminals for households might seem to be totally different than those designed for management reporting, in reality they are a subclass of the latter. The home market has no massive requirements for text and data input, which is a relatively minor addition to the massive reporting needs that are close to what the British call Viewdata plus.

To create an image of the type of equipment which is required, it is advisable to return to the Prestel consumer's interactive TV set. The answer page permits the consumer to do buying through the system. A formatted page is provided, and to it the user inputs data with a simple numeric keypad. Typically, the answers involve a multiple choice. For instance, the message pages offered by Fintel present the user with the following:

0 Just called to say hello!
1 Please call me.
2 Send me information about Fintel.
3. Would like advice on using Fintel.

Cursor positioning within this limited message capability is quite efficient. The cursor jumps lines when the box key is pressed. When the cursor is in the line, the user keys in the chosen number. An example is ordering out of a wine list. (However, the interactive capabilities of an in-house system are far broader than those in the BPO case. When operating within a controlled environment, one can offer so much more to the user.)

Let us conclude by saying that terminal design is no longer the easy job it used to be. From videocolor and graphics capabilities to more complex tasks such as editing and framing screenfuls, the terminal's job has changed in character and function. Present technology opens up a totally different perspective of requirements.

Terminal Configurations

There are a number of ways in which terminals can be designed to operate from the proposed PDI codes. The two most likely ways result in terminals which incorporate either character-oriented or bit-map display memories. Because the PDI codes are intended to be terminal-independent, they must be interpreted before the contents of the two different display memories are generated. For that reason, it is most likely that all future videotex terminals will contain a microprocessor to perform the interpretation and generate an appropriate display memory code.

Based on that most likely approach to the design of terminals, there are a number of possible configurations. Their exact nature will depend on the features manufacturers think are important to the sale of their products after considering the cost and complexity of incorporating those features in the design.

One manufacturer may aim its product at the individual consumer market and limit the display capabilities in order to sell at the lowest possible price. Another may design business terminals capable of displaying, or printing, business forms with shaded rectangular areas and 80 characters to a line. Still another may think there is a market for terminals among hi-fi fans who are willing to pay for high-resolution graphics display.

Terminals incorporating alphageometric display memories are capable of displaying images with much higher resolution than terminals with character-oriented memories; but they require more memory elements, and that increases their cost. However, the cost of implementing terminals with bit-map memories capable of displaying high-resolution images must be balanced against the undesirable features of character-oriented displays with their inherent limitations.

Other design considerations also are important. For instance, larger input buffers could store several pages of a document transmitted as a single burst of information instead of one page at a time with the communications lines held, but not used, as subscribers peruse each page.

The PDI will interpret incoming instructions and generate the appropriate code to be inserted in the display memory. The way in which the instructions are implemented will depend not only on whether the terminal is bit- or character-oriented but on the features and resolution the terminal is capable of displaying. One terminal may simply display all images as black on a white background, or vice versa, whereas another, more complex terminal may provide both colored and gray-scale images, as well as features to "blink" individual objects in the final display.

The processing is relatively simple and can be easily performed with an 8-bit microcomputer—much better with 16-bit machines. In a character-oriented display terminal, the interpretation of commands would be similar, but the hardware or software display generator would be very much different.

Character-oriented terminals are suitable for low- to medium-resolution graphic images. In those terminals, the display memories are relatively small and contain codes which represent alphanumeric characters or mosaic symbols to be displayed in fixed positions. A system of that type is best suited for textual messages with a fixed format because the character shapes and locations are predetermined and can be coded efficiently. The displayed output is, however, rather limited for graphic images in character-oriented terminals containing memories with only a single 8-bit word to describe each character position displayed on the screen. One limitation is the terminal's ability to change from one color to another as additional memory space is required to encode the color change.

A further consequence of using only one 8-bit word for each character position is that some words must be recognized as command words, which reduces the number of bits used to address the read-only memories containing alphanumeric or mosaic characters. The remaining 6 or 7 bits can address only 64 or 128 different symbols; and in the graphic mode of operation, the mosaic characters are typically constrained to combinations of dots in a 3 by 2 dot matrix. This provides a resolution of 60 by 80 elements on a 20 by 40 character display.

Both of the limitations can, however, be partially solved by extending the length of each word contained in the display memory to more than 8 bits. Extra bits in a word of greater length could indicate background and foreground color changes for each character position. Other bits could extend the resolution of graphic images by providing addressing for 256 different graphic characters defined in a 4 by 2 dot matrix. This would provide a final resolution of 80 by 80 picture elements for graphic images, but the individual spots would not have square outlines if the 3:4 aspect ratio of the television screen were maintained.

In contrast, terminals incorporating bit-map display memories are capable of much higher resolution images than character-oriented terminals. In bit-map terminals each spot of picture element displayed on the screen has a corresponding position in the display memory. Thus, for a terminal with a resolution of 320 by 240 picture elements a memory containing 76,800 positions is called for. It could be built from charge-coupled device (CCD) memory chips now available.

The possibility of keeping the options for future developments open is of prime importance. Pictures described by PDI means allow for both forward and backward compatibility; that is, terminals need not immediately become obsolete if a system is upgraded to provide more precise data for specialized terminals. Older terminals merely interpret data in the best way they can, and newer terminals are not constrained by the design criteria of older terminals.

Terminals should be designed to accommodate more powerful graphic-modifier instructions to permit rotation, scaling, and transposition of portions of the displayed image. Instructions for generating images might be called picture manipulation instructions (PMIs), and will be needed by IPs to generate the images for storage in VDBs. These instructions would allow IPs, or others with appropriate terminals, to generate and manipulate their own images by using interactive devices.

The best solution, then, is to use a CEPT-PLP compatible color terminal which supports business graphics and charts through bit-level coding, makes possible a spectrum of 21 colors from a 64-color set (though no more than three colors in use at any time is advisable), gives a good-quality image presentation, and is designed for interactive charting.

For graphical purposes, at the user's choice, the type of presentation should be decimal, log-decimal, or log-log, with different modes being supported: trend, trend and area, venn, circular, histogram, and 3-D histogram. Gantt charts are another example. Graphics should be automatically prepared for the user through resident software, and the same thing is true of desired color changes. Such facilities enhance a user-friendly environment because they make it possible to translate data and tables into graphs without the user having to intervene in the mechanics of a graphics presentation.

Such features support the four topmost considerations in regard to *executive productivity:*

1. The ability to communicate information through electronic managing

2. The ability to integrate on the same intelligent workstation (WS) diverse functions spread on different mainframes and specialized nonintelligent terminals

3. The facility to experiment on financial and operational planning with easy-to-use tools and present the results in an interactive manner

4. The ease in reading public databases online, which is one of the pillars of the knowledge society

Highly paid men and women must work with their heads to produce knowledge. Handwork in the office and the factory can best be done by robots. To produce knowledge, managers and professionals should steadily upgrade their skills, methods, and equipment. One of those improvements is videotex.

Implementing a Videotex System

*"The sooner you start coding your program,
the longer it is going to take."*
                    H. LEDGARD

Chapter 4 started with electronic mail and ended with videotex to emphasize that the latter service is a more sophisticated follow-up of the former. With advances in technology, the optimal configuration for any given application will continually change. As this evolution rolls on, videotex will move into an intensive use of personal computers. Not only are the two not competitive with one another, they are complementary.

- The mid to late 1970s saw electronic mail offerings as an improved version of telex effected with public telephone lines.

- In 1979 Prestel (the British Viewdata system) went public with considerable excitement about its possibilities, and the pace accelerated.

- In 1980 the need for communicating databases brought forward Bildschirmtext (the German videotex version).

- In 1981 the less than 64K, 8-bit processor, Apple microcomputer caught the attention of many prospective buyers who thought quite highly of its processing power.

- In 1982 the IBM personal computer received much notoriety for reason of its enlarged memory capacity and the 16-bit microprocessor.

- By 1983 the Apple II and III and the original IBM PC were overshadowed by products offering bigger, better, faster, and less expensive capabilities, including a hard disk, 2 MB of central memory, and 32-bit microprocessors.

Two major groups have been targeted for videotex applications: business and home users. To business, videotex is primarily a database-data communication medium serving information needs that are measured in a small number of hours. In this sense it fits between realtime computers, telex and other wire services, and printed media. Electronic messaging, online entry and retrieval, and fast analysis of business events are examples of business implementation.

For home users, current videotex offerings are based on entertainment, shopping, home banking, reservations, computer-based instruction, and so on. For home as well as business, pilot tests and field tests have dominated the scene, with a few commercial operations appearing only in the last few years.

Precisely because videotex is not only a recent development but also one in full evolution, future directions cannot be reliably predicted by past trends. From system design to market offerings, solutions must be based on a vision of expected videotex activity. Management should ask and obtain a realistic appraisal of the prospects for videotex in the near future, including the range of applications and the role of graphics. The latter is a direct reflection of the fact that videotex is a communications system that offers, in addition to the text and data capability, pictorial presentation and a repertoire of display attributes.

Text, data, and pictures are intended to be displayed by using television raster standards of the different countries. As we will see in the following chapters, alphamosaic, alphageometric, and alphaphotographic displays may be used, though data for each type of display is separated into different presentation protocol data units (PPDUs) during transmission.

Evident Advantages

Videotex has evident advantages, and not all of them have been put in perspective so far. It is therefore proper, before we proceed with the mechanics of the public service, to highlight where the benefits lie and why the process deserves attention.

Figure 6-1 shows the online system capabilities as they compare with the now classical realtime systems (such as those for banks and airlines) and the evolving videotex solutions. With videotex emphasis is on:

1. Simplicity
2. Ease of use
3. User-friendly environments
4. Interactive capabilities

5. Easy-to-learn procedures

6. Low-cost computer resources

7. Low-cost terminals

8. Inexpensive, standard software

9. Videocolor features

10. Low-cost graphics

11. A base in existing, proved technology, not in developments yet to come

12. An open door to the extended use of bit-level protocols with image rotation, videophone, teleconferencing, and other services

Between the classical solution and the original Viewdata approach (as developed by Sam Fedida at the BPO) there is, of course, a facilities range. The gap will be closing because of the steady drive to reduce complexity and cost at the classical realtime site and the growth in videotex sophistication such as provided by the CEPT-PLP protocol. The steady stream of improvements will change the overall picture of

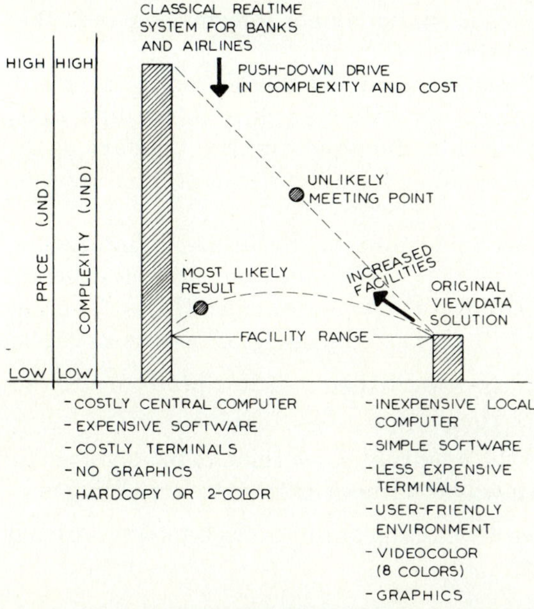

Figure 6-1 The original Viewdata solution was far removed from classical realtime. The service gap is closing, but the videotex costs remain very low compared with the costs of traditional approaches.

both public and in-house videotex services as offered today. For instance, the possibility of moving a page back and forth, the capability of sending telex-type messages, and the integrating of the videotex image with low-scan TV will open new vistas.

Color and graphics are key features of videotex; they increase the challenge of designing readable pages. That has been demonstrated in England. Many IPs, when they first sit at a Prestel keyboard, appear to be driven to use the full range wherever they can, often arbitrarily and to the detriment of the final result.

With some experience, designers have begun to abandon the rainbow approach. Many of the most experienced IPs now suggest the wisdom of:

- Using very few colors

- Standing by a proven store of color combinations, often only two or three

- Using the other capabilities seldom if at all

Color choice is largely a matter of taste and effect on the viewer; it is affected to a greater or lesser degree by the make of set and the lighting conditions for viewing. These are important considerations not only because of their definite impact on videotex but also because color and graphics are the big new issues in management information and other interactive communication systems.

In their experience with Prestel, many IPs discovered, by accident, that different makes of terminals not only reproduce colors differently but also vary in the aspect ratio of the page and the typeface of the character generator. As we all know so well, there is no standardization in terminal design.

Applications design is most important to the future of interactive systems, and pending applications in the direct line between classical realtime and videotex will enlarge the original capabilities. The key issues here are:

1. To preserve and maintain the established videotex protocol (character level)

2. To keep the door open for the eventual conversion to bit-level protocols (at least through gateways) without major system changes

3. To observe and support the developing database technology, which is applications-independent

If these prerequisites are properly supported, the increased sophistication in the way videotex works will not result in higher cost figures. It will not, that is, provided that local intelligence is enhanced through economic solutions such as those offered by the new generation of micro-computers, the optical storage media, the intelligent plastic cards, the communicating copiers, and so on.

That is why, in Fig. 6-1, a clear distinction is made between the "unlikely meeting point," which lies about halfway between the very expensive classical realtime solutions and the original low-cost view-data approach. Given the right choices at the implementation stage, the most likely result will be near classical realtime in terms of range of facilities but also near the original Viewdata solution in regard to the very important issue of costs. We will return to this subject.

An Implementation Timetable

Through videotex the database-data communications technology is spreading throughout the organization and creating an environment in which text and data access is not only easy to the end user but also secure. This is quite important as industry, business, and financial institutions find it increasingly necessary to obtain diversified information which should be made accessible to an ever-increasing number of users.

A steadily growing population of end users can get significant advantages from transparent, yet effective, access to information which through very simple commands can be retrieved and even structured to the end user's specific needs. The databases accessible through videotex contain corporate assets whose employment is vital to the functions of management. But videotex also makes it feasible to reach public databases in an economic and speedy manner.

The terminals themselves are of low cost—the house TV set with or without microcomputer support—and the operating instructions are very simple. Videotex permits wider compatibility among terminal devices than the now-growing PC implementation. Owens-Corning Fiberglass, for instance, has observed that after several IBM personal computers had been acquired, primarily because of their increased central memory over the Apple II and the possibility of doing 3270 emulation, incompatibilities entered the system. Software like Information Master, which runs on the Apple II disk operating system, could not run on the IBM, and so on.

On the other hand, experiments with the universal interface supported by the videotex protocol have had commendable results. Participants found it relatively easy to learn to use the videotex terminals in a couple of hands-on hours, and there were no problems in changing the nonintelligent terminal hardware. (However, in many cases it took up to 2 days to become really proficient. The amount of time required for users to learn the electronic mail routines ranged from 2 to 6 h.)

Experience also was that equipment service calls were infrequent and the participants felt they had much better control over computer uptime. Very few hardware problems were encountered and since the videotex software is well past the baby-sitting stage, no software ones at all. Logical security was assured through closed user groups (CUG). That is particularly important because many information providers intend to offer their services either to a specified, external population (subscribers) or join the public videotex service only to have available a low-cost, internal communications system.

We have said that videotex can be viewed as an improved, more sophisticated electronic mail offering. By supporting mailboxes, it permits the transmission, reception, and exchange of messages among users. To be connected to the videotex service, the end user must be assigned a code number. Once this is done he can receive and send messages through his mailbox. The system automatically flashes the code or name of the sender (given one of the two inputs) and asks for the code of the receiver(s). After the code is input through a numeric keypad, the name of the receiver is written on the videoscreen by the computer. Now assured that everything is in order, the sender has at his disposition 20 lines (out of 24) of 40 characters each to insert his message.

The operations which have been described make evident the simplicity of the installation and implementation of such a system. A typical timetable, starting after management's decision to proceed with an experimental implementation (and to put up the corresponding budget), is as follows. *The preparatory work will take between 1 and 2 months*. It includes:

- Deciding on the group of users to be taken on
- Making a tentative menu selection tree (for the experiment)
- Establishing the routing pages and designing the infopages as functions of the selected population and the application to be chosen

The 1- to 2-month time period presupposes two qualified people working full time on the project. It is highly inadvisable to start with a larger systems group. As the popular saying goes, "Nine women will not make a baby in one month." Adding more systems specialists will delay the project rather than advance it.

The training of the end user on how to employ the videotex terminal should not take more than one day. Earlier, mention was made of a couple of hours of hands-on experience. In preparation for that, however-er, it is a good idea to have a one-day seminar which introduces the prospective users of videotex services to the overall concept. The seminar should be conducted with small groups (some 10 to 15 people) and cover both the dynamics and the mechanics. The hands-on experience should follow.

The risk of failure is too high if the organization does not provide the prospective users with the necessary new concepts. Let me repeat: To assure good utilization and lasting success, the best approach is to spend a day on training the users in the fundamentals and then spend a couple of hours per person on hands-on experience at the desk. Supposing that the starting group will include about 20 to 25 executives and/or clerical support, the two specialists assigned to the project can conduct both the seminar and the hands-on experience in a week. That is so much better because it avoids diluting energies and forgetting what the users have already learned.

There is also time to be devoted to *the input of information to the viewdatabase* through an editing terminal, as well as the bulk loading of the VDB to the mainframe. This introduces the need for videotex utilities. A number of utility programs are necessary to interconnect the central computing facilities to the public or private videotex network. The most important are:

- A videotex control program
- An editing facility
- A communicating monitor
- A terminal manager

In regard to editing, it is inadvisable to write conversion software which will typically extract text and data from mainframe or mini files, format it in infopages, and load it onto the VDB. There are packages available to do that job. However, for the sake of the sporty system designers who would like to write their own, the main lines of a project set up to that end are described in a separate section.

At the same time, it is advisable to avoid manual data input through an editing terminal on other than an exception basis to fit in special infopages to be retrieved by the end user. On the other hand, the use of an edit procedure will be a normal practice with electronic mail and when any text, data, or graphs are broadcast workstation to workstation through videotex.

Prototyping

We have spoken of beginning the videotex introduction with a user group of 20 to 25 people and enlarging it as internal experience accumulates. The use of a pilot group, or *prototyping,* is a valuable way to develop videotex applications. Though it adds to the cost and somewhat delays the wider spread of videotex services, in the long run prototyping gives the user a real feeling of involvement and becomes a time-saver by helping to debug the projected system.

In other terms, prototyping both fleshes out the critical dimensions of the problem and increases user satisfaction. When we prototype an application, we learn its parameters and know what resources we have. Given that we have started by putting the proper tools in place, chances are that the prototype will tell us how long it should take to develop the system. Through prototyping, we really create a sense of control and ownership on the part of users and also system specialists, though for different reasons.

One of the key advantages of a pilot project is that the project enables management to establish a measure of effectiveness:

1. Determine if videotex in its current offering can provide an effective alternative to such other presentation media as the classical computer terminal

2. Gain experience with videotex and user acceptance that would lead to proper policies and procedures for videotex use

3. Provide immediate support to areas of clearly identified need whose payback or priority has been inadequate to justify immediate custom programming and the commitment of computer resources

4. Develop an internal prototyping capability in support of the development of more sophisticated systems

For the purpose of evaluating the success of a pilot project, objectives should first be established and then categorized as user-related and technology-related. Among the user-related objectives are:

- Immediate response capability
- Time savings through the response capability
- Timeliness of reported text and data
- Improved quality of information at the executive desk
- Ease of use in information retrieval due to both menu selection and direct access
- Prompting and help pages available online

- System availability and reliability
- A range of communications and database access capabilities

To overcome acceptance problems, it is wise to follow a strategy that establishes and maintains user enthusiasm, contains costs and timetables, emphasizes packaged software, and slightly leads the existing, proven technology but also focuses on the basic facilities to be offered the user.

Users are much happier with the new information technology than in the past, and the reason is simple: involvement. We start involving our users in the processes of:

- Improving the performance of our key systems
- Setting priority assignments for new development efforts
- Planning for the future by defining new concerns

Involvement and participation have actually helped to turn around the user relationship with the systems specialists and the managers of computer operations.

Videotex can be of significant importance in this process because a good part of keeping users happy is rapid dissemination of the data they request. Through an interactive videotex approach, the end user gets a highly visible program for a modest investment of effort. A big plus is that the approach promotes a dialog between computer management and the end users.

The technical factors include the utilization of advanced technology, the ability to do prototyping, valid security measures which are generally supported, assurance about data integrity, and the possibility of experimentation which leads to development alternatives prior to final choice of the methodology to be employed, the specific services to be supported, and the user population to which videotex will appeal.

A good example along this line is the First Bank System's offering videotex services to a user population among its clientele in a farming area of the Dakotas. The basic objectives set by FBS of Minneapolis include: agrimarketing (commodity prices and market commentary), agriproduction; shopping (national, regional, local), financial (banking and economic analysis), home and family, local happenings, and what's new, as well as weather and sports. The most valuable add-ons are enterprise management (transaction data capture starting with bill payment and decision models) and market strategies contributed by established companies in the field.

Figure 6-2 shows three typical infopages from the FBS offering. The videotex pilot objectives have been to study consumer reaction to home

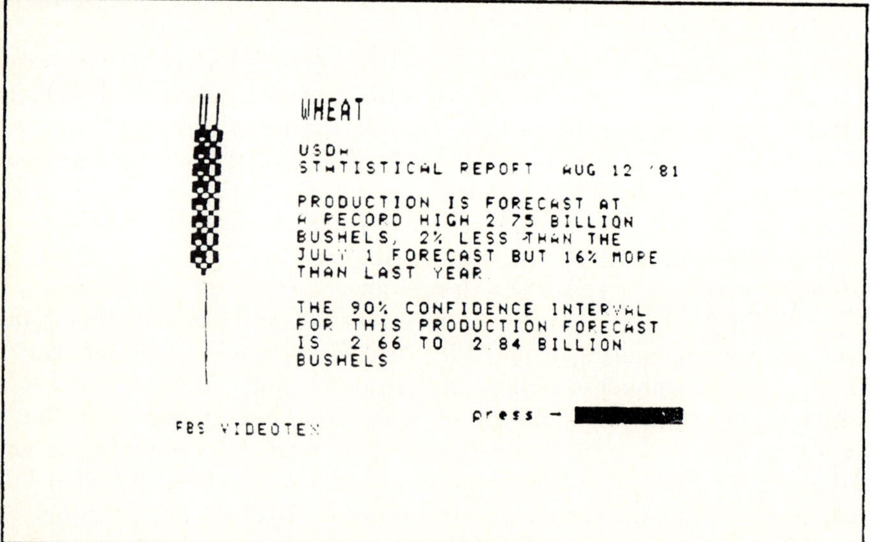

Figure 6-2 Three typical infopages from the videotex offering of the First Bank System.

delivery, evaluate the demand for specific services, provide documented evidence on consumer willingness to pay for the service, and assure an acceptable level of security and privacy.

Let us now see what the findings have been. The first is that home banking cannot stand alone. The system has a basic cost, and home banking does not provide sufficient income to pay it. Banks classically work through paper. To the consumer, paper handling is sufficiently convenient and, most importantly, inexpensive (though it is expensive for the bank).

Yet the videotex experiment also documented that, if the price is right, banking services will be accepted as part of a home information system. Other requisites are that the information be targeted to specific needs and be timely, reliable, and locally oriented. Typically, such information will be supplied by entities interested in offering services to consumers. Among the IPs are Agridata Resources, North Dakota State University, the First Computer Corporation, First System Agencies, Fargo Forum, Valley City *Times Record*, Wahpeton *Daily News*, J.C. Penney, Dayton's, B. Dalton, Lutheran Hospitals & Home Society, St. Paul Fire & Marine Insurance Companies, Pillsbury, and, of course, the financial institutions of the First Bank System.

This significant range of IPs helps demonstrate that vendors will pay for access to videotex subscribers, sponsor information, and, through it,

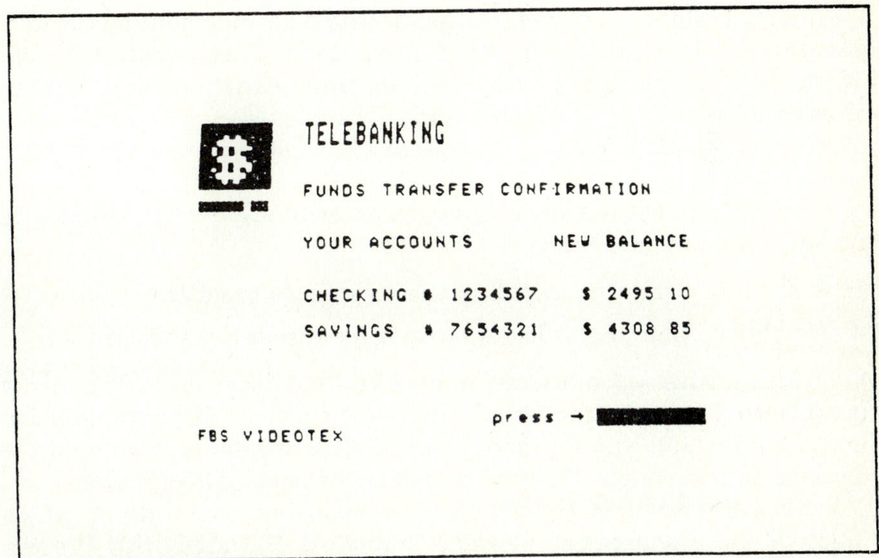

Figure 6-2 continued

sell their products. On the end user side, home delivery via videotex and computer was well received. Technology did not frighten the consumer. In the Dakota experiment, farmers were willing to pay for a business tool. Their concern over security was minimized through the use of personal identification numbers and the chip in card for scrambling and as a record of transactions.

At the same time, user reaction documented that system reliability is essential to establish product quality and credibility. Videotex is an integral part of the bank's delivery system that complements and uses other forms of electronic fund transfers. The consumer likes the service; and the greatest barrier to home banking will be marketing, not the computer-communications system.

Videotex, the First Bank System has learned from experience, has marketing potential today. Some 67 percent of the users said they would get videotex: early, 29 percent, and later, 38 percent. National research has supported those responses. Farmers are willing to pay $300 for the terminal and $30 per month for the service. Reaction will, however, be much more positive if costs are $5 to $7 per month—as they should be—and are kept at that level.

A Videotex Budget

Even after the prototype run has produced valid results and document-ed the wisdom of going ahead with videotex implementation, a lot of questions straight from the traditional system development methodolo-gies remain to be answered. What about the cost? How is it justified? What about compatibility with the mainframe? Are there any addition-al costs to provide it? Similar questions can be asked about operating systems, network communication protocols, capacity, and maintenance responsibilities.

When we talk of the videotex budget, an evaluation must necessarily include two alternatives:

1. A private system installed on the company's premises
2. A publicly offered service to be paid for through a subscription

Each alternative will put a terminal on the end user's desk. The cost of the terminal is not a function of whether the private or public solution is chosen; instead it is a function of the presentation protocol and the decision to use simple TV-type terminals or personal computers.

In the 1979 to 1984 timeframe, European implementations which followed the alphamosaic viewdata standard, initiated with Prestel, have been able to use the common television receiver coupled with a decoder and modem attachment. That has been true of Prestel in Eng-land, Bildschirmtext in Germany, Videotext in Switzerland and Spain,

Videotel in Italy, Viditel in Holland, Teledata in Norway and Denmark, Telset in Finland, and Datavision in Sweden.

Charges have varied from country to country because the videotex services have been run as monopolies by the PTTs (postal, telephone, and telegraph services—government utilities). The best example of ingenious establishment of charges to promote the videotex service was in Germany, where the PTT (Bundespost) applied:

- A DM 5 (about $2) per month charge for decoder and modem
- The cost of a simple telephone call for a connection to the VDB

This works out to 24 pfennigs (about 10 cents). It is levied once per connection independently of time in Berlin, and in West Germany it is charged for each 8 min. of connection time. PTTs in other countries apply the basic charge to each 3 min. of connection time.

That picture, however, changed radically with the adoption of the CEPT-PLP standard. As is true of the LAN (IEEE 802 Committee effort), the implementation of a standard must be paid for in hard cash. In the case of videotex this means purchasing a new television set with alphageometric capabilities. In Germany, the TV costs DM 2000 and a decoder another DM 1000 for a total of DM 3000, or about $1250.

Such a high cost per end user cannot and will not be paid by everybody, and the effect of the CEPT-PLP standard will be to slow down implementation by dampening enthusiasm for videotex use. A study recently made by Germany's major financial institution, the Bank für Gemeinwirtschaft, demonstrated that although professionals will pay the cost necessary to avail themselves of videotex service, particularly for home banking reasons, other citizens will not.

That picture may, however, change in the coming years. First, interfacing personal computers to the network will cost less than interfacing unintelligent TV terminals. Second, there is always the possibility of developing an interface device on a chip. The Japanese, for instance, are now talking of offering an interface at DM 450 (less than $200).

Other pricing mechanisms do, of course, exist. An example is the joint videotex venture by Centel (the fifth largest U.S. telephone company), Field Enterprises (newspapers and broadcasting), and Honeywell offered in Chicago in mid-1983 at a price, including the videotex terminal, of about $25 a month.

Now let us look at the cost of the central resource. *To the end user* the cost is transparent. In many cases, as in home banking, access to the database costs nothing. There is, however, a charge for some infopages, particularly those offered by service companies. The latter cost is variable by page and IP. It usually amounts to a few cents, and the user is informed about the cost of the page by an annotation in the top right-hand corner. Typically, the public utility offering videotex collects the money charged per page, on a factoring basis, on behalf of the IP.

To the information provider the charges can be of two different types. On centralized systems, where databasing is done exclusively on centrally installed mainframes belonging to the utility, the IP must pay a rental for the pages occupied plus connection time. On distributed systems, it is connection time and the switching facility which have to be paid for. In both cases, the public utilities (particularly PTTs) have dual tariff systems for day and night hours. The latter costs are about a third of the former.

If the organization decides to hire its own computer, it has three alternatives:

1. It can do so by timesharing a mainframe. This is the least advisable approach.

2. It can dedicate a mini. That was the preferred solution until recently.

3. It can make videotex available through local area networks and workstations. That is the best solution for the future.

Some LAN offerings include videotex; examples are the Nestar/Zynar Cluster/One on Apple II. This possibility can be expected not only to continue but to expand, and a broad range of systems supporting videotex at the PC level in the coming years can be foreseen.

The costs associated with a minicomputer solution tend to exceed $100,000 for hardware and software. One system which I employed reached three times that level, mostly because of the software (videotex) cost. As the user population expands, those cost figures should be reduced significantly. At the microcomputer level, the cost experience is no different than the one we know from systems such as those for database management as (DBMS). With the Nestar/Zynar LAN it is enough to allocate one PC to run the videotex programs in a network containing several multifunctional PCs.

An evident advantage is of a functional nature. It is much better to share applications on one terminal than to place at or around the executive's desk a number of terminals each oriented to a specific job and handling only one function. *The proliferation of terminals leads to confusion, and it should be avoided.*

In financial terms, the solution I suggest will cost less than $2000 for the dedicated PC and about $1500 for the software. The total cost will be about 3 percent of the cost of the minicomputer solution. There is no cost for the terminals, because they are multifunctional and videotex is an add-on. Without doubt this is the approach to be followed.

Designing the Applications Environment

"It is not the lofty sails, but the unseen wind that moves the ship."

W. MACNEILE DIXON

Videotex has been designed to free users from dependence upon a complex communications environment and to provide a memo-and-letter distribution facility which can employ standard channels. This is an evolution in online practices largely dictated by experience: In the beginning of realtime, the service typically allowed attaching simple terminals to the host.

The resulting facility was little more than a restricted file copy mechanism. Little or no specialized software was provided to aid in the creation, projection, and final disposition of messages. However, the advent of packet switching networks gave a boost to new concepts which were both the origin and the result of a demand for improvements in function. At about the same time, there was a switch from private to public networks: Common-carrier-oriented, network-based data communication and message systems began to appear although initially they were intended for connecting user terminals to each other and to mainframes.

Slowly, we worked toward support of data communication and word processing on the same system, the use of color, voice store and forward, the connecting of thousands of point-of-sale (POS) terminals and automated teller machines (ATM) on the same network, the wider use of portable terminals, voice data entry, and the use of graphics in the output.

Videotex supports color, graphics, data, and text, and developments are now underway to incorporate a voice capability. And while those facilities were being implemented, the awareness about their need opened up new avenues that led to business opportunities for some companies, hardship for others, and departures for still others which know how to use the capabilities that technology makes possible.

The newspaper publishers who reacted negatively to AT&T's entry into the videotex market did not miss the opportunity to use the capabilities the process presents to improve their own product's marketability. In late 1980, while most subscribers to the Columbus *Dispatch* were reading the newspaper in their usual way, they could also tune the same news in on their home computer screens. The transmission marked the debut of one of America's first commercial electronic newspapers. The *Dispatch* (an afternoon daily with a circulation of 200,000) began distributing its news, along with Associated Press and United Press International wire service stories, through the CompuServe computer information network.

At the last count, CompuServe had 30,000 computer subscribers, 8 percent of them in the Columbus area. From 6 p.m. to 5 a.m. weekdays, weekends, and holidays the contents of the *Dispatch*, except for such features as comics and ads, are stored in a central computer waiting for CompuServe subscribers to beckon them to their computer screens. Subscribers can also call up general information about computers, video games, and programs for making financial calculations.

Using the Screenfuls

The new mode of online text, data, and image communications is in its infancy. This means that the existing environment is still quite limited in size and many potential communicants are not yet attached. But videotex is a healthy counterweight to the profusion of networks, and it guarantees the ability to send messages to any other user. Indeed, its basic objective is the handling of messages, which are not limited to text but can include drawings, facsimile, and the like. The process has a structured segment containing some header information, and it has an unstructured body of text.

A user may have a program to help in the acquisition of the message parts and then initiate transmission. Delivery of messages to users is quite rapid. The recipient has available a system which aids in processing messages for viewing and disposition, including filing or deletion. A new approach to message communication is thus evolving. It is based on videotex characteristics as a modernization of the classical realtime concept, but it is a substitute for telex, business mail, memos, and perhaps newspapers.

One way to look at videotex (and its electronic mail capabilities) is as a worldwide compatible service like telex but one designed for a higher data signaling rate, of 2.4 kbps upward based on the extended character set of an editing typewriter. Its terminals would be suitable for use with a public utility, and they would replace common electronic typewriters with editing and possibly processing capabilities.

This is, however, just one example. In substituting for classical real-time, the ways and means of conversion to videotex are more at the user's choice, but here again standards have to be followed. It is understood that the potential of this promising network architecture cannot be realized without an appropriate set of communications protocols.

That is not unlike the great variety of requirements to be met in the field of teleprocessing at large, which resulted in an increasing number of system constraints. Any additional requirements imposed on these services can be met appropriately by more structured approaches. In this sense, the formalisms implied by videotex are but an extension of a trend.

Still, in spite of formalisms, videotex is more flexible than classical realtime:

- It is expandable to support more users than RT service.

- It accommodates a flexible design of infopages to meet the projected usage.

Some aspects of the formal approach to videotex communications are shown in Fig. 7-1. The screenful design reflects either static information pages intended to inform the user but on which no alteration can be made during the inquiry (hence, of a *read only* nature) or response frames. The latter would differ in respect to the means the user has at his disposition for data (or text) input:

- Numeric keyboard only

- Alphanumeric keyboard (with the response frames providing limited fields)

- Practically blank pages available to the user

The latter presuppose, at the user's end, the availability of an alphanumeric keyboard, special functions, graphics terminals, or memory-to-memory supports. This makes possible a decentralized organization in which videotex frames are used as a high-class telex facility between units.

To use the videotex capabilities the best way, the user must first complete a reasonably good description of what the problem is. This will involve an answer to such questions as:

1. Total database volume

2. Update volume

3. Update frequency

4. Response time required

5. Topology of the network

As with all computer applications, a clear distinction must be made between format and data. The format will include headings, clear text, references, and other matter. The data will change from time to time but not with the same frequency. From that comes the division at the drafting board between static and dynamic pages suggested in Fig. 7-2 and, further, of the dynamic pages themselves between static and dynamic fields. Each has its own screenful design requirements.

Standards reflect a self-evident need. When local, incompatible message protocols are used, special conversion programs are required to transform local messages to a common format and to retransform messages to local format when they come into the local environment. Such transformations potentially lead to errors, delays, costs, and information loss.

The videotex message format provides features to capture all the information any local message system might use. It also assures a reliable communication procedure by guaranteeing a transport-level protocol which conventionally provides calls to open and close connec-

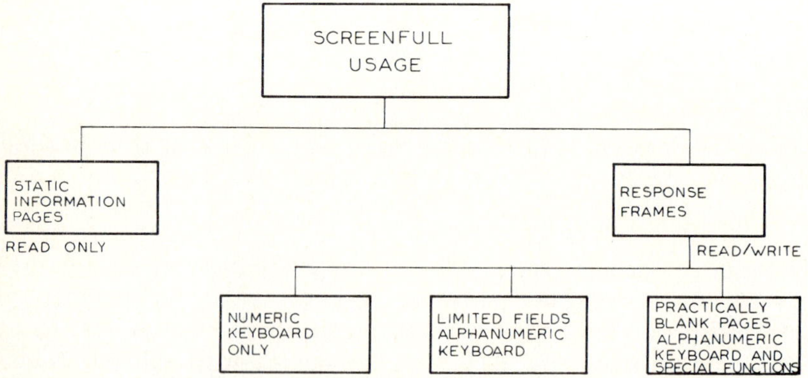

Figure 7-1 A hierarchical, formal approach to videotex communications through interactive pages (screenfuls).

tions or send and receive data on a connection and means to signal and be notified of conditions at the local level. Since this service uses the voice grade lines reaching every home or business office, it requires a set of interfaces.

Handling Messages

The original videotex system was based on ready-made frames linked together through a tree structure. These were report- and action-oriented, and the software provided an interactive editor. As applications expanded, new possibilities have come into view. One example is that of the Columbus *Dispatch* and CompuServe mentioned in the introduction to this chapter.

The user hooks his telephone receiver into a device that connects his computer to CompuServe's. After he types out his identification number for billing purposes and a password, words begin appearing on his screen at the rate of 30 letters a second, slightly faster than the average person can read: "Welcome to CompuServe." The machine then begins to list CompuServe's offerings, which include news, weather, and sports. Having selected news, the user is given a choice of headlines and so on down to the detailed news.

Price is, of course, a problem, but a subscriber has the option of buying a $30 per month packet, sold at Tandy Corporations Radio

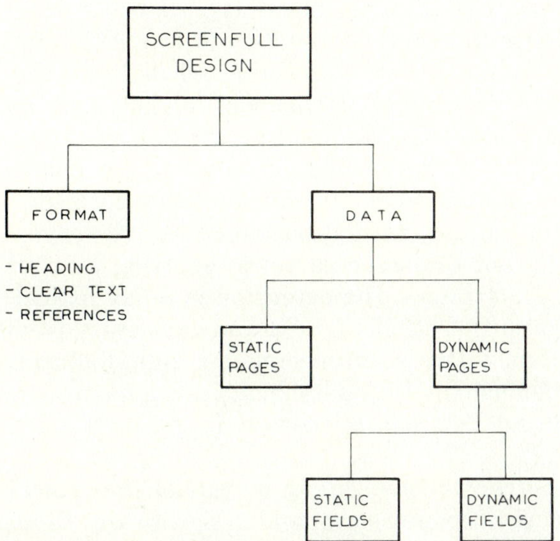

Figure 7-2 Distinction made in screenful design between format and static and dynamic pages.

Shack stores, that provides the hardware and software necessary to use the CompuServe system. And prices are falling. Several vendors will be selling hardware and software packages able to turn any television set into a terminal for under $300.

On the information provider's side, going electronic was easy for the *Dispatch*. Its reporters and editors already worked on video display terminals tied into a central computer. The only conversion required was a program for the computer and a copy editor to shorten headlines, separate stories into categories, such as local news or sports, and rank the stories by importance.

Since videotex is not a monolithic solution, several offerings have been made, starting with the appropriate market test. An example is the venture of Knight-Ridder Newspapers, Inc., which spent a rumored $1.5 million on a 6-month experimental offering of its Viewtron system in Coral Gables, Florida. News came from *The Miami Herald,* a Knight-Ridder newspaper; Dow Jones & Co.; Barron's; *The New York Times,* and the Associated Press. Other publications, such as *Consumer Reports,* fed selected articles into the system, and there have been several advertisers.

Dow Jones has been using computers to transmit news stories and financial information since the mid-1970s. About 13,000 terminals, mostly in brokerage firms, tie into its two-way system, which is transmitted over telephone lines. The charges are $50 a month plus $40 an hour for office use (brokers pay a wholesale fee) or $3 a minute for the first 3 min and 50 cents/min thereafter for home use.

The Dow Jones News/Retrieval service provides stock market quotations, company profiles, financial news, and selected articles from Dow Jones publications. Furthermore, a cheaper version (lacking Boolean search for *The Wall Street Journal* itself) is available in three areas in the United States on two-way cable for only $22 per month plus $9 monthly rental for equipment if the end user does not own a micro. That fee is for unlimited use, a great bargain.

Management reporting is not the same as news reporting but, for those who have made the experiment, the extension is fairly logical. Another exciting area of potential benefit is text and data entry from fairly remote places. Redifon, a British minicomputer manufacturer, promotes the idea that we can use the Viewdata facilities augmented by compiler to obtain "VD Plus." This is a misnomer. It is a data entry interactive system.

Given that roughly 30 percent of the computer costs of the typical company arise from data entry, Redifon's proposal is significant. Properly structured, data entry can evolve toward a videotex-like solution and in the process change into a text and data entry capability:

1. Stand-alone microcomputer-based videotex terminals (most likely with bubble memory) will communicate through a publicly supported (telephone) network and mainly perform text and data entry.

2. Through a communicating database (pioneered by Bildschirmtext in Germany and now reexported to England) such videotex terminals will communicate both with mainframe databases and with other management-decision-oriented videotex systems.

3. Telephone companies will provide the necessary network architecture and its gateways, and the user organization will look after the *rephrasing* necessary within its own operations (mainframes, minis, micros, and videotex hardware).

Quelle, a German Sears-like chain, is moving in that direction, and Barclays, in England, seems to be also. Most likely, the three listed functions will be the constituent parts of a successful videotex system, but the time is not yet here. We have to prepare for it, and the best way is to start with the bread-and-butter management reporting structure.

Simplicity as a Basic Goal

The design of the applications environment should follow some principles which have been well established by experience with practical, user-oriented installations. They are 10 in number.

1. *Keep the system simple.* As soon as the system becomes complicated, it might as well be run on a larger computer at greater expense and with longer lead times. It is poor practice to confuse the primary goal of videotex—fast, easy-to-handle management reporting—with other worthwhile but different functions such as controlled data entry. The requirements are too diverse to group on the same machine—unless our goal is inefficiency.

A private videotex system gives the using firm total control over its operation, and it provides facilities for storing or retrieving information essential to the running of the business. But videotex is not designed for volume data capture. Even though the response frame facility does allow data entry, it cannot compete against a conventional solution for program development or highly transactional realtime operations. Videotex does not challenge the existing systems designed to meet those needs. It is as different from conventional computing as telex or word processing is. It is a new tool with a whole new range of possibilities: a system designed to provide fast, accurate, up-to-the-minute visual data to a widespread population.

2. *Spend time and thought on the structure of the information to be accessible for management purposes.* To be effective, we should try to

consider all eventualities when we start out, allow more space rather than less (frame numbers), and create a VDB such that we will not have to revamp it because we have run out of space, the hierarchical tree was not designed for possible expansion, or the original design was too myopic.

Projected applications, which may take some time to materialize, should be outlined at the outset. At the Vickers subsidiary Howson-Algraphy, the first application has been the competitor's products database. TV screens offering access to the database are located in several company departments and in the boardroom. Later the application will be extended to the homes of executives via their own domestic TV sets.

The management services department also uses the system to report progress on computer projects for user departments, together with personnel information, manning levels, and vacancies, available computer facilities and their costs, and systems performance statistics.

3. *Set the rules and conventions very early in the design phase.* The rules do not need to be universal among videotex installations; they should define the standards to be observed in *your* organization: Examples are color codes and the types of information and text associated with the different user departments. The same thing is true of graphs.

Response time, cost, and ease of use are among the videotex assets. The terminals are based on TV sets; they employ a well-understood technology and are made in large quantities. That means they can be supplied with a degree of economy no other terminal can approach. Also, we are talking about a multipurpose terminal that is able to receive normal entertainment programs and, at no extra cost, provide access to databases. Such a terminal is not constrained by the computer mainframe we may have, and it is a fairly standardized unit. Standardization also means that implementation of a system is dramatically quicker and easier than normal computer network implementation, and it can be undertaken by less skilled personnel.

In a simple, effective approach requiring no special skills, videotex can provide information that is accurate, easily updatable and highly visual to a wide audience of people at a reasonable price. And, as with the telephone, the user can have access to a wide area.

4. *Keep the man-machine interaction user-friendly.* The system should be easy to operate at all times. To be so, it should force the user through a path. The less the user is obliged to depress keys the better. Complex protocols should not be used.

To be forgiving, the software must be very resilient; it should not take the system down if the user makes an error. Rather, it should inform the user of the error. "Sorry,..." This is quite important because most end users are inexperienced with computers and videotex is, after all, computer-based.

5. *Keep the VDB isolated from the mainframe.* The transfer of pages (text, data, and graphs) from the mainframe database to the VDB should represent a small fraction of the former's volume. If it gets large, we face the wrong problem: We replicate the mainframe files on the mini, and we simply do not need two levels of databases.

Also, if text and data are sent from the mainframe to the VDB in the same form they are available on the maxicomputer, we are wasting our time. Quite similarly, if we increase the number of applications, programs, and masks, use of a mini for the VDB becomes unwise.

6. *To take advantage of an independent VDB facility, treat the videotex system as a separate exercise. A good test is "Can I prepare the VDB frames manually?"* Though one of the goals is to use the computer power in extracting and formatting from the big database, the manual test is a good starting point. Having made it, we are in a good position to evaluate the wisdom of online editing.

Another criterion of independence from mainframe operations is that of taking a classical computer output, emulating its conversion to VDB, and applying a rule to evaluate its possible use. First, how many copies are to be made? If many, it is a good videotex candidate: Rather than many hard copies, one soft copy is available to everybody. Second, is paper going through another process prior to reaching top management for decision?

- Is it consolidated manually?

- Are exceptions extracted from it?

- Is some other operation that we do not always know about "necessary"?

Is somebody going to edit or amend that data prior to passing it on? If yes, then it is a good candidate for videotex. Third, is paper which is out of date as soon as it is printed going around the organization? Think of:

- Telephone directory

- Scheduling directives

- Product definitions

- Motivation schemes

- Price book

From an evaluation standpoint, such questions are most important because the overriding demand should be that the cost of what is done is justified. How much does it cost now? How much it will cost by videotex?

7. *To experiment with and learn about the requirements the VDB should meet, walk through the total projected application.* Within a banking environment, for instance, a videotex system will give its user

access to competitive information, pricing of services, contracts and policy, product specifications, details of meetings, minutes of meetings, and briefings on potential prospects. System anwers will greatly depend on the kind of facilities we require from the system: electronic mail, realtime update, response frames, editing, accounting, and so on.

Users with the mailbox facility may leave messages for other users. Such messages are entered through a normal editing terminal and stored in a dedicated file. For broadcasting, messages may be sent from the system console to a selected user or to all users. Memos can be left in the system so that, on coming online, the person for whom they are intended will be told who called and whom to call. When staff spend time out of the office and need information on the state of the business at any given time, a videotex system can make available to them any information they need.

8. *Since graphics can ably assist management in decision making, graphical presentation should definitely be part of the preparatory tasks.* But remember that the object of management graphs should be to give direction; no fine print is needed as in engineering design. Editing should be concerned with graphical information. Forecasts against plan, projections, and trends should be presented in graphical form. To merge text and graphs, graph page integration should use the facilities supported by the videotex system. For instance:

- Assign the first page to a summary graph.

- Assign the next single-digit pages to detailed or associated graphs.

- Assign the two-digit pages to supporting tables.

Graphs are also very important because the best way to introduce videotex to top management is to show results, and graphs are eloquent. But they do not need to be three-dimensional. An adapted time-sharing routine may suffice for a computer-based graphics preparation. The organization of the graphics frames into a comprehensive sequence (Fig. 7-3) is more important to management than an alphageometric design. Typically, a timesharing routine will consist of a source program in Fortran able to extract data in tabular form and prepare curves, histograms, and bar charts. What is necessary is to study the videotex of presentation. Is the output convertible to a form?

Other relevant questions are these: Is it possible to introduce color? For daily curves, is one color acceptable? For exceptional cases, can graphs be set on a different medium, for instance, the editing terminal? Is integration of the graphic frames feasible with the basic editing package?

9. *In reworking a given mainframe (microform or paper-based) application, take advantage of the facilities offered by the videotex system.* A

case in point is the electronic mail capability. The mailbox facility enables one user to send messages to another. The user must be registered with the group mailbox to have access to this facility, which basically uses two pages: 940 to send mail and 950 to receive mail.

After page 940 is called, the short name of the addressee, or names of the addressees, must be entered up to 14 names at any one time. Next to each short name, the system will display the corresponding full name. When short names have been entered, each terminated by pressing #, a blank page with a mailbox header will be displayed, and upon it a message may be typed. Full use can be made of the cursor controls. When the message is complete, the user should press interlock and end on the edit keyboard to send it. A message sent to a user is first sent as a *new message*. After accessing the message, the user has an option to *delete* or *store*. If the message is stored, it can be referred to or deleted later. Messages can be broadcast from the system console to a single user or all users. It can be a 40-character message or a full-page broadcast.

A user who requires the facility to receive messages must be a member of the mailbox user group and have an individual short name. Messages awaiting the attention of the user will be found on page 950. A note will automatically inform the user of messages waiting on that page, and the note will be displayed on the welcome screen after dial-up.

10. *Finally, always remember that videotex uses the most widespread facilities in existence: telephones and TV sets. System design should take advantage of that fact.* An individual user can send messages to other

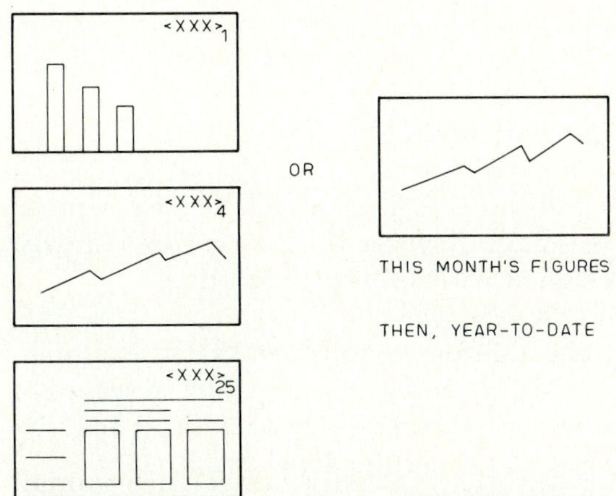

Figure 7-3 Particular attention should be paid to the organization of graphics pages in the viewdatabase. Here is a multipage organization that, by pressing #, assures movement to the next page.

users just by knowing the names of the users with whom he wishes to correspond. All that is necessary is to format the message and send it to the system. The recipient will be automatically notified that a message is waiting and be told to access his message pad (a specific page). If the recipient is not currently using the system, he will be informed of the waiting message as soon as he dials up the system.

Some Advice

If a banker visits a client's office to discuss a loan and does not have the information with him, he can dial the videotex system and get the details (provided they are stored in the VDB.) Through the response frame facility, information can be entered back into the system from the calling television set.

Memos and circular letters to the branch offices are another broad applications area. If we consider what it currently costs to prepare, type, duplicate, and distribute the information currently sent throughout organizations, we find that videotex is not only a faster and better way but also a cheaper way to do it.

Although the guidelines presented here can help in the design of an efficient videotex system, it is important to keep in mind a list of things that should be carefully avoided. For instance:

1. *Don't undertake too much too soon. It is better to be selective.* Design the system on a limited basis, but establish a broad perspective. Then, select the priorities and the applications that are critical to system success. This approach to design is a major difference between classical DP and videotex.

2. *Don't emulate the classical, paper-bound report presentation.* Don't make the videotex forms in the image of paper records (printed output). Details, if support, should be oriented only toward day-to-day operations. Management should get only highlights and exceptions and be protected through security and clearance.

3. *Don't mix DP with VDB and end user functions (EUF).* Remember that videotex is not data processing. It serves the user-oriented layer, and the frames should be designed with that goal in mind.

4. *Don't mix different types of applications on the same machine.* Different types of applications have different requirements. For example, management reports and massive data entry are way apart in regard to the requirements they place on an interactive system. Both may be handled by videotex processes, but not on the same machine.

However, "exception" text and data entry—electronic mail, for example — can be handled by the same machine that processes management queries. Note that we are talking about exceptions.

5. *Don't forget to provide on-call service.* The end user (particularly at the management level) can be frustrated in certain cases (time out and routing) without baby-sitting. Anything which might discourage user interactivity should be avoided.

6. *Don't overcrowd the pages.* Whether they present data, text, or image, the pages must be easily readable and their information understood; 100 words or fewer is plenty. Time Inc. thinks 60 is about right. The same thing is true about using color.

7. *Don't change the format of a page unless it is absolutely necessary.* Changing formats disorients the user and may lead to mistakes. Therefore, to avoid unstable formats it is necessary to look carefully at the human engineering aspects of page design prior to final implementation. But once the decision on implementation has been made, it is advisable to stick to it.

8. *Don't use too many terminals on the same machine; they will degrade response time.* Remember that the primary goal of videotex is to increase the mental productivity of the user. Anything which obscures this goal or deviates from it should be avoided.

9. *Don't fail to incorporate decision data into the system.* Videotex should be primarily oriented toward senior management rather than duplicate what is done with classical DP. This, however, calls for a thorough analysis of text, data, and image requirements by management and for management. The design guidelines both follow and answer these requirements. Videotex should present not detailed data, but only exceptions, summaries (with up to three significant digits), and graphs.

10. *Don't fail to use what is available in data assets.* Currently available data resources (out of classical DP) must be used prior to a search for new data sources; the system design effort should be concentrated in using what is available more efficiently. Only then can the full power of videotex be realized. Another pitfall is to avoid putting priorities on the implementation: "Let's decide later on it."

Finally, because one of the key assets of this system is the user-friendly environment which it supports, a substantial number of pages will be dedicated to routing. Hence, *don't* underestimate the volume of the routing frames. They tend to be half of the pages in use.

8

Planning for Videotex

*"Those who cannot remember the past are
condemned to repeat it."*
GEORGE SANTAYANA

A general appreciation of the role videotex can play on a relative scale
(high, low) involving the number of simultaneous users and the information capability made available to those users is given by Fig. 8-1.

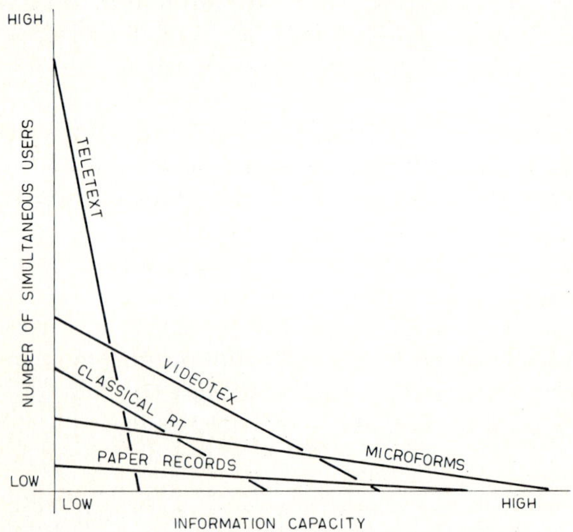

Figure 8-1 An order-of-magnitude comparison of videotex,
teletext, classical realtime, microforms, and paper records
based on information capacity and the number of simultaneous users. Teletext and videotex are complementary
rather than competitive in nature.

The illustration contrasts videotex with teletext, classical realtime applications, paper records, and microforms.

As can be seen in the graphical presentation, microforms have definite advantages in terms of both parameters over paper records, while being more compact, easy to access, and less costly overall. The same factors characterize the main competitive advantages of videotex over classical realtime. Table 8-1 puts this contrast in better perspective. Teletex and microforms present advantages and disadvantages as compared to videotex, but these three modern office management tools are complementary to rather than competitive with one another.

A fundamental point in these comparisons is the projected ability to master the uncontrollable mass of paper every office is snowed under. Certain statistics show, for example, that we need 1.5 tons of paper to produce 1 ton of iron! But that benefit requires both new physical media and brains to work on system studies (logical support).

For example, work is now proceeding on *picturebases*. Analog images will be microform-stored on digital optical disks [Video long play (VLP)], and digital optical recording will be used on the outer tracks of the disk for indexing and retrieval purposes along videotex lines. This DORVLP hybrid, which is currently under development, seems to offer considerable advantages to the office user. DOR (digital optical recordings) is well suited for videotex because the latter is slow-moving information. Also, with the DOR solution information never gets lost because, given the nature of optical disk recording, we cannot overlay.

Table 8-1   Videotex versus Realtime

Classical realtime	Videotex
Realtime update, but also specially designed, hence expensive, software	Real enough time supporting general use capability
Direct system effort oriented to software design	Off-the-shelf videotex software
Involved systems work with large databases	Bypass of large DB-oriented concern
Limited resources left to work at the end user function (EUF) level	No applications programming requirements (extending the life of aging software)
Few if any resources left to work at the EUF level	Possibility to allocate resources to VDB design and EUF
Typewriter-like devices at the user level	Simple, user-friendly terminal (and protocol)
Mistakes can be serious	The system forgives the user's errors

Software Facilities

Fairly advanced text, image, and data management functions are expected to become operational in 1984 or 1985, but current videotex technology makes broad applications feasible. In a banking environment (at both headquarters and branch offices) are services offered by videotex center on information which has always been on paper. Videotex eases the recording, retrieval, storage, and distribution problems. In an industrial environment, videotex is well suited to inventory control, the tracking of spare parts, personnel management, the administration of sales orders, property listings, and the like.

In any case, a basic preoccupation is the steady upgrading of the videotex supported functions. The basic software is in full evolution and now includes:

- The standard operating system

- A run time support package

- Utility programs

- A system database

- An applications database

The system can display a list of the special functions available and another list of the normal access commands, help correct erroneous input, and interrupt output with a command.

Special software for text editing supports:

- Cursor manipulation

- The transposition of characters and graphics to other frames

- Recall of predefined graphic fonts

- Insertion and deletion of rows and columns

- Insertion and deletion of characters and graphics

- Insertion and deletion of frames within information pages

Utility programs both assist the end user and facilitate system management by providing creation and modification of different profiles, the capability of accounting management, and (for IPs) bulk editing capability. Not all commercially available videotex systems assure the same facilities, but practically all are in a process of steady development. Typical routines include the updating of tables and distribution lists, along with a store and forward capability.

Electronic mail is usually handled as a matter of course, but at varying levels of sophistication. One public videotex service currently in development will support three types of mail—third class, normal, and registered—and will also inform the sender whether a message has or has not been delivered. Specific functions by class of mail are outlined in Table 8-2.

As with all computer projects, great attention must be paid to the prerequisites, the thorough systems study. In designing a business system, *safety* comes first. The second key issue is *simplicity*. Videotex conforms to the entry and exit conventions in a user-friendly manner, but great care should be exercised in database design. The available software sees to it that the database administration function is helped by such routines as database creation, database copy, database compaction, and database-structure printout.

But the design phase cannot be escaped. Some of the available videotex software supports a system DB and a user DB. The overall design reflects experience with large commercial databases and aims to make feasible usage by the layman (push button) if the proper preparatory work is done.

Database design should, therefore, be a primary consideration, and the work program must take due account of that fact. Figure 8-2 demonstrates that VDB design must start immediately (at the time of ordering the hardware) and be a couple of months advanced by the time the necessary "unload" facility from the mainframe is developed. The object of the latter is to see to it that, from the spool of the batch programs running on the mainframe, we can perform automatically three distinct, well-defined operations:

1. *Extract* (this is applications-dependent on current batch output) and preformat to VDB protocol

2. *Format,* which involves an applications-independent manner
    21. Edit (fine programmatic interfaces)
    22. Structure of the infopage/table presentation
    23. Color
    24. Graphics
3. *Load* operation at mainframe level

Table 8-2   Electronic Mail Facilities

Third class:
  Read and destroy
  Destroy without reading
  Overlay

Normal:
  Read
  Register on paper
  Store on computer support

Registered:
  Return a receipt to the sender identifying delivery time and intended user

Primary emphasis should be placed, in the first phase, on the development of an extracting facility able to provide a tabular form of presentation including color for exception. This is the wise thing to do, and attention must be focused on it. In the second phase, graphics, which can originally be prepared through an editing terminal, should also be converted to an online application.

The subject of converting data stored on a mainframe to the VDB is so important that we will devote a whole chapter to it. Though as a beginning experience we can comfortably prepare the report frames manually through an editing terminal, it is unthinkable to do so when the users start demanding a steady stream of data and graphs. We will be returning in detail to this subject.

To summarize what was said in this section, a videotex system rests on a number of software- supported facilities. These have to be learned by the user's specialists in detail prior to planning the system's implementation. Other important facilities of which the specialist must be cognizant are hardware-based. We will take a look at them prior to continuing with the planning premises.

Videotex Decoders

The decoder permits the TV set to be connected to the telephone network for communication with a database. It can be built into existing

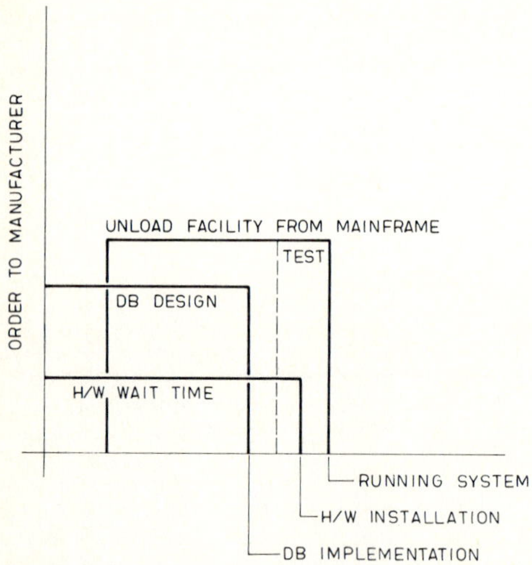

Figure 8-2 With a valid work program, viewdatabase design should start immediately after ordering the hardware. Shortly thereafter the building of the interface between VDB and mainframe should begin.

TV sets or into adapters for attachment to existing units. A decoder is composed of five devices:

1. A *line isolator*. Its function is to protect the telephone lines from high voltages present in the TV.

2. A *modem* (modulator/demodulator) to convert the analog telephone signals into digital and the digital into analog.

3. A *memory* capability to store data for display.

4. A *display generator* able to transform characters stored in the memory to dot patterns.

5. An *input processor*. Its function is to control and synchronize component operations, process incoming signals, and store the signals in memory.

Components 3, 4, and 5 are common to videotex and teletext decoders; but instead of a line isolator and modem, a teletext decoder needs a *frame grabber* which serves as front end to the TV set (Fig. 8-3). In a few years, technology will radically change the shape of the devices we currently use. Another important component is the data input unit at the user site:

- A decimal keypad

- An alphanumeric keyboard

- Magnetic supports

- Eventually, voice input

The solution provided must be such that additional peripherals can be added easily and with the components interchangeable. An upgrading

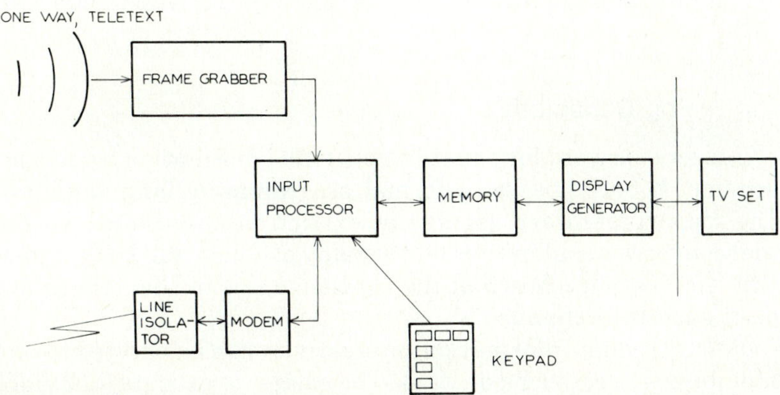

Figure 8-3 Blackbox organization of videotex and teletext from source to destination.

of current videotex input would be personal and small-business computers able to receive programs loaded downline from videotex centers. Information providers will typically use editing terminals to create and maintain videotex pages. These differ from subscriber terminals in having special facilities for entering color and graphic information such as the creation of lines, circles, rectangles, and squares.

Online editing terminals operate under the direct control of the VDB and are intended for the preparation or modification of limited quantities of information. They incorporate a color video display unit capable of presenting the full videotex character set and an alphanumeric keyboard with additional keys for cursor manipulation, color and graphics control, and special functions such as flash and double height.

Offline editing terminals have facilities for local information storage which permit batch transmission of formatted material. These are basically word processors equipped with a color video display unit supporting the videotex character set and control information such as keywords and routing instructions. Among the word processing functions available are addition and deletion of text, indentation, formatting, and global replacement.

Domestic sets usually include remote control of output because a cable connection for the keypad is awkward to use. An additional feature is a separate function button to retrieve the main index page, dial alternative videotex centers, and autodial or autoidentify and go back to the preceding frame.

Small screens designed for close-up viewing are more appropriate for business, particularly if they are placed on the desk by the telephone. They are monochromatic to keep the cost down and also because it is expensive to get adequate resolution in color below a 13-in (inch) screen. Business applications will, furthermore, call for printing capability and for bulk storage media. A multifunctional set design is shown in Fig. 8-4.

Establishing Dialog Capabilities

"Dialog" is a generic word for a preplanned man-information communication scheme. It encompasses a formal programming language, programming support for interrogating the text and database, and a host of less formal conversational interchanges many of which will be designed for specific applications. Much of the challenge lies in the able establishment of such interchanges.

In principle, the man-information interactivity can be effected in one or some combination of five basic ways. The easiest to execute (but more complex to remember) is the one-shot request to text and database

(TDB). This can be either *direct* (meaning that the communicator will identify to the machine, in an absolute code, the information element sought) or *mnemonic*. The second possibility is a keyword to be converted by the equipment into a direct address.

In the general case, the information element sought has no immediate relevance to the method which will be used, though in special cases the choice of the latter might be influenced by the former. Most important, this independence is sustained in management-type applications in which the amount of information sought is actually limited to a few pages containing data vital to making a decision.

Here the highest priority should be given to the user-friendly features of the method to be adopted. That brings into view the third method: *menu selection*. It is a solution to information search by which a number of *routing pages* are examined until the infopages sought can be

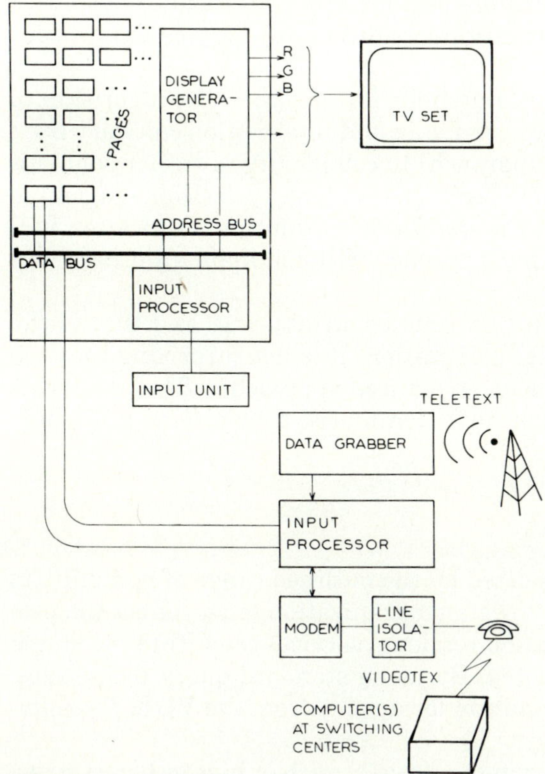

Figure 8-4 A multifunctional set design with memory capabilities for videotex applications.

located. In that sense, menu selection is a sequential-type tree structure and therefore a hierarchical solution. We will elaborate on it with specific examples because it is the key to the VDB.

The fourth approach is to use *query languages,* which are usually offered by computer manufacturers or as packages written by software firms for a specific system. They allow much more freedom than menu selection because they do not have to follow a treelike (hence, inflexible) route. However, they also demand much more knowledge on behalf of the communicator (the person using them), and so they are better suited to information specialists than to the average person with the problem.

Efforts to blend the last two classes have been made with variable success which is usually limited to applications-oriented solutions. Such efforts have led to a fifth alternative: *special man-information communication languages designed by profession* or by job within a profession. An example is the powerful foreign exchange language which allows the Forex operators of a leading Swiss bank to compose images on their screens quite similarly to the way they calculate with pencil and paper.

This much can be said in a nutshell: The choice of the algorithm to guide the communication between man and information can vary from simplicity (a user-friendly approach) to complexity (a rather sophisticated presentation) and from specialization (one profession, one job) to generalization. Eventually, the man-information communication will be in spoken English and that approach will dominate all others, but the time is not yet ripe for it.

A simple, generalized solution implies structuring, and that is the price to pay for assuring user orientation. It is not surprising that the videotex solutions capitalize on structured approaches. That contrasts sharply with practices which have dominated the systems profession from the beginning.

Facilities with Videotex

Let us now see what is supported by the outlined range of possibilities with the understanding that system organization must reflect the projected applications. Information retrieval services constitute the single largest application for videotex. By using standard query techniques, subscribers can select information for display from the VDB. Two simple methods are utilized:

Selection by alternative through a tree (branch or hierarchical) structure. At each level in the hierarchy a subject is divided into further detail. The user, guided by a few simple instructions, selects from the numbered topics by merely keying the digit corresponding to the wanted item. Selection is repeated until the desired level of information is reached.

Direct access. Each page in videotex is identified by at least one keyword: its unique page number. Pages may also be identified by one or more alphanumeric keywords, and information retrieval capabilities are thereby enhanced.

Messages between subscribers are usually fixed-format. They can be completed by using only a simple numeric keypad to select from variables. Free-format messages can be created with the assistance of an alphanumeric keyboard. Messages are transmitted in store and forward. Direct mail IPs can utilize the file qualification capabilities of the system for direct mailings to specific subscriber groups (narrowcasting).

In the conversational mode, a subscriber may be able to "dial" another subscriber and enter into a direct dialog conversation by visual display. Page browsing is another capability which enhances videotex usage for mail-order and holiday (consumer travel) organizations. Connection to telex also is possible, with telex messages transmitted directly by a videotex subscriber to one or more telex terminals. Transaction services can be as varied as placing a reservation, requesting a sales visit or purchasing goods via credit card.

As time goes on and experience is gained, videotex systems will evolve as assemblies of one or more mini- and microcomputers and their associated peripherals operating with videotex software. A typical system might include:

- Mini- or microprocessor
- Remote diagnostic and hardware maintenance capabilities
- Disk drives for optical storage systems
- A number of ports for the connection of terminals and interconnecting other videotex systems

The subject that is most critical to the user (and to the future of videotex itself) is access to the VDB. The access can be direct by a subscriber or be machine-actuated as in the case of communicating databases. The latter will be discussed in a subsequent chapter.

There has been considerable controversy over what exactly videotex software is — an OS, a DMBS, a data communication facility, or something new and strange. Let us see why the primary goal (and service supported) is a functional enhancement of communications and interrogation capabilities and therefore is suitable for the new output concept.

Interrogation is the selection of data from a text and database followed by extraction (essentially copying) of what is selected either for intermediate processing (such as sorting, adding, summarizing, and formatting) or for presenting readable reports to the end user. Information is usually entered to identify the part of the database on which the interrogation is to be performed. This is followed by two statements: one of the selection criterion and another of the action of extracting. Identifying the part of the database could, in effect, be considered part of the selection criterion, even if practice sees to it that it is given separate treatment.

As a self-contained system, videotex treats interrogation (and hence dialog capabilities) as an action grouping in which interfile relations are possible. The complexities of the operation are totally transparent to the user. The resulting action initiates a page transfer from memory to visualization, while the host language is reduced to a single number no longer than the one we use for a typical station-to-station telephone call.

In this way we can call up text or graphics: the result can be a routing page (Fig. 8-5A) or an infopage (Fig. 8-5B). It may contain primarily data (Fig. 8-5B), a message (text) quite similar to a telex (TWX) service (Fig. 8-5C), or graphics (Fig. 8-5D). It can be in black and white but is usually in color. Note also that at the upper-right corner each frame (page) carries a number. This is the direct address, an alternative to the routing capability we will examine later.

Obtaining a Better Service

The concept of the new output services dominates the discussion in this section. The text, data, and graphics that videotex handles (voice is scheduled to be added at a later time) are stored in memory for ready access. Inquiry will be used for a variety of purposes; an example is sales and promotion analysis. Promotion executives can use videotex in a commercial or banking environment to develop new customers, handle the current base, provide for base extension, arrange visits, or consult statistics and reports. Videotex can present summary data on the following:

- Business activities
- Client actions or reactions
- Regional results
- Branch office and salesman statistics

Text and data handling can have extensions well beyond the classical data processing or text processing functions. Videotex is a potential computer-aided design (CAD) tool, and it is likely to have an important impact on CAD. (We will return to this subject toward the end of the book).

By harnessing a new communications technology, Videotex opens up the designer's options. It will provide vital links between different companies and between distributed centers and thereby give designers the basic information which is the lifeblood of their work. There is little doubt that IPs will shortly appear in the marketplace to provide access to databases containing design information.

The concept of staff members distributed at different physical locations and working together in such a way that they can collectively form an operating unit may well soon become a reality. Physically distributed designers will form one of the centers of competence likely to emerge. Potential applications include progress reporting between

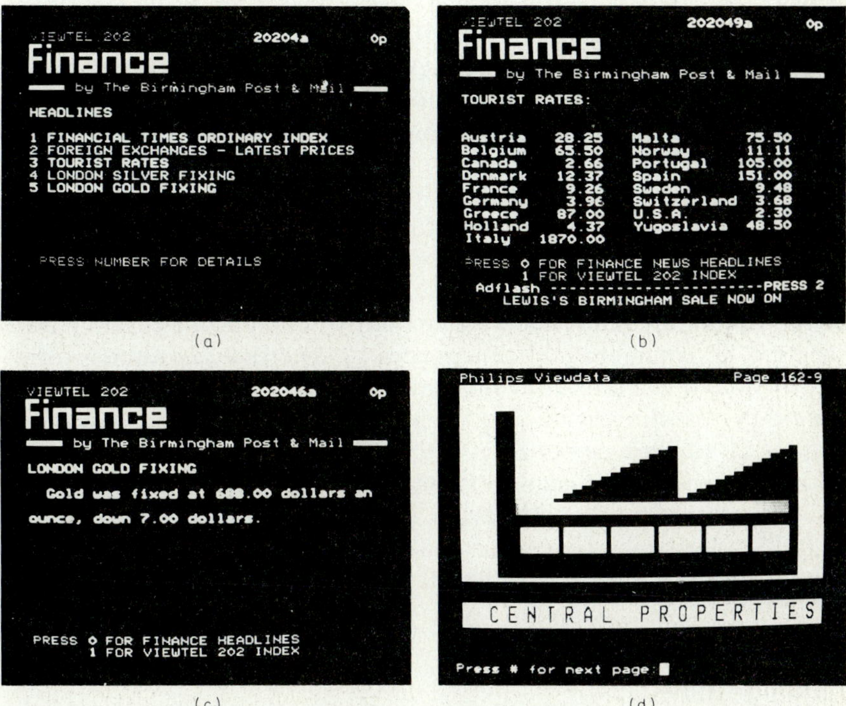

Figure 8-5 Examples of videotex pages. (*a*) Routing; (*b*) primarily data; (*c*) a message similar to TWX; (*d*) a graphics presentation.

subcontractors and main contractors in other engineering sectors such as the mechanical and electrical fields.

Some facilities necessary to support this environment are available in the current systems. For instance, a number of function commands are incorporated; they permit the user to redisplay the current frame immediately. (This is useful when corruption of data on the telephone line has rendered the displayed frame invalid or suspect.) Retrieval of the latest version of the current frame is possible, and this is useful when the page or frame being accessed is realtime updated on the database. Other useful functions that have been implemented include retrieval of the last three frames accessed one by one and in an order the reverse of that of the original retrieval. Branching outward on any of the options presented is possible from each of the frames retrieved by using this function. And the same thing is true of the cancellation of a miskeyed command.

These features make it feasible, in addition to passing information to work forces, to report progress back to an administrative center. Videotex is a potentially powerful management tool for reporting work done, project status, and exceptional events. Because of the telephone's widespread use (both cable- and radio-supported), videotex is useful in connection with distributed projects and mobile teams and situations in which the work of subcontractors must be carefully monitored. It permits an urgent query to be addressed to a location linked by telephone for progress control and the expediting of delayed supplies.

As stated earlier, several developments such as new or adapted graphics protocols are still needed so that graphical information can be transferred effectively. For rather bulky data transfers, it will be necessary to use data compression equipment to maximize efficiency. Furthermore, in commercially or technically sensitive situations, terminals will have to maintain information security because videotex, by sharing the existing voice-frequency telephone network, is a working example of an open system.

But the preceding discussion is not intended to imply that videotex is the best system for all applications. In fact, it is not. Today the process fits halfway between batch and realtime systems. It is a real enough time system, but it is also interactive.

Real Enough Time

Two issues are dominant in a discussion of real enough time (RET):

1. Database management
2. Driving of output systems

Database organization and structure are independent of the different output media which will be used. And that is true of the information pages no matter which system is adopted. In other data processing approaches, different output media require different solutions. This is not true of videotex.

The environment which videotex supports is such that the file of information is *not* related to the output to be obtained on softcopy. Five or ten years ago it might have been made to fit batch requirements. The indexing algorithm, which, as we will see, interfaces between the old structure and the presentation of the data to the user, serves a purpose conceptually similar to that of virtual store. Both the purpose and the mechanics are different, however. (Note that this kind of interfacing will be very helpful in the years ahead because of the galloping technology. It can, for example, be foreseen that the same discussion—of layered independence—will take place in the late 1980s because of the advent of videodisks.)

Since online inquiry through videotex implies a tree structure and such a structure rests on firm organization, it follows that, to create the form of the search tree, we should first decide on:

- The sort of data we wish to manage

- The choices of branching (going up the tree and backwards)

For instance, the now classical approach in a public videotex service is to present in the first frame the choices:

1. Business information

2. Financial-economic data

3. Personnel

4. Technology

5. Organization

6. Marketing

7. Laws and rules

8. Import and export

Another key choice is the amount of use the system will have. For how many people is it projected? For a large-scale videotex information retrieval system to operate economically, it has to deal with many simultaneous users. To assess the capability of the computer to support a number of simultaneous users, assumptions were advanced regarding:

- The requests per minute a user was likely to make in the information retrieval mode

■ The size of the data page in terms of the number of characters which a computer response might involve

Since the time of the original studies, two approaches have been considered. In the first, the computer would poll and service the users in turn. Any spare time remaining would be used for housekeeping duties or to support interactive services not related to information retrieval, such as calculations and messages. This introduced the complications of interleaving a polling algorithm, involving many unnecessary computer inquiries, with random scheduling. The alternative approach would be to use a wholly interrupt-based mechanism with a fast hardware microprogram. Other studies were conducted to facilitate the information retrieval process.

Even if retrieval (particularly distributed retrieval capability) is mastered, the fact remains that videotex is intended not for realtime databases, but for RET. Unless data updated as of one hour or one day ago is acceptable, videotex should not be chosen as the medium.

More precisely, if a large amount of information changes very much

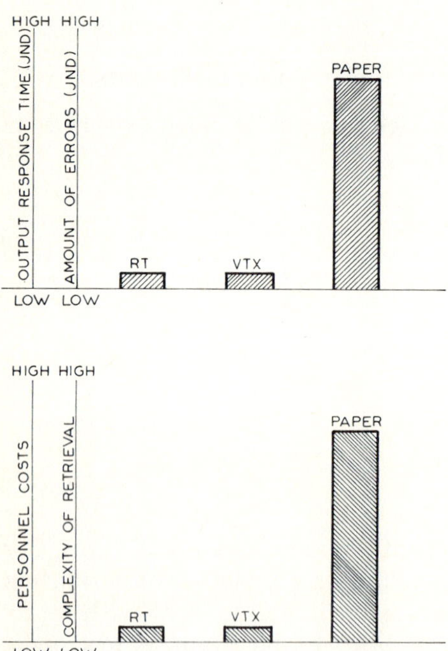

Figure 8-6 A comparison (just notice difference, JND) of realtime, videotex, and paper based on output response time, number of errors, personnel costs, and complexity of retrieval.

or very little over a given time period, videotex approaches are not advisable. For very much change, the classical realtime is the means of choice. For very little change, the answer is paper or microfiche.

Then, why videotex? Because it has its strong points. It is a low-cost, user-friendly, interactive medium which makes it possible to bring together a unique selection of information elements and access them through means not previously available. It is a reasonably low cost approach to an online alternative to paper handling: *The information is made available to those who want it and with computer precision and timeliness.*

That eliminates two types of problems: errors and delays (Fig. 8-6). The frames shown on videotex are computer-supported. If the database is free of errors, so the online presentation will be. [The scale is just note difference (JND).] A gain in productivity has to be realized by automating document handling and distribution, and that is what both videotex and realtime can do. Personnel costs are reduced in comparison with manual paper shuffling (which justifies the higher equipment costs), and the complexity of the retrieval job also is reduced. In terms of

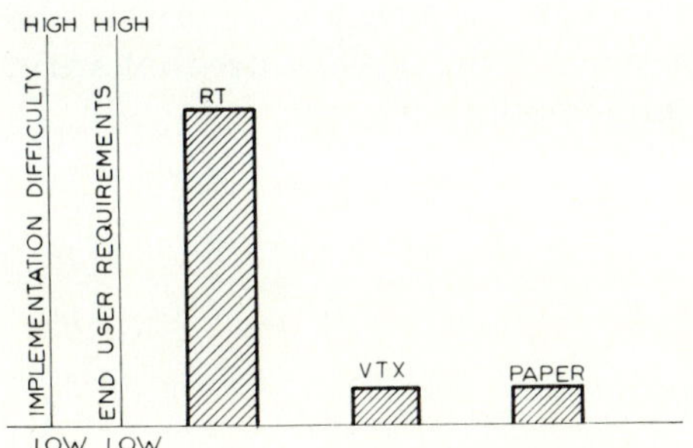

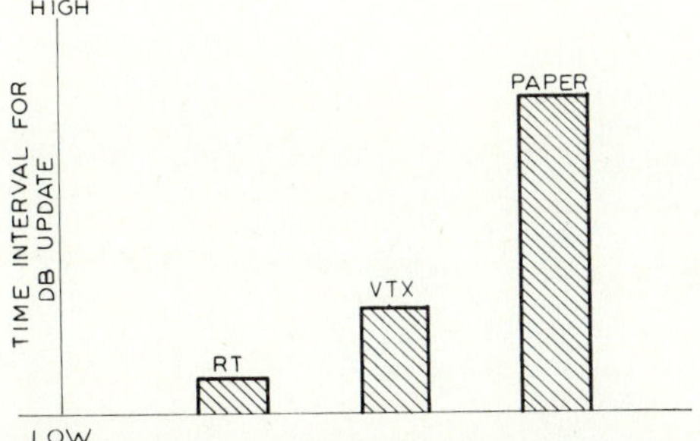

Figure 8-7 A comparison of the three factors of Fig. 8-6 based on implementation difficulty, end user requirements, and the time interval for database update.

response time, the process is far superior to paper shuffling and equal to realtime access. Updating is, however, slower than RT, as we have already stated.

The output of videotex is typically visual, but nothing forbids adding appropriate gear to provide hardcopy: a typewriter (less likely), facsimile (more likely), microfiche, microfilm, phototypeset, or some other type of printing. All that is needed is to send the output to the user, and that can be easily done. In fact, decisions on how to handle the output should be situational, and in making them both technology and costs should be taken into consideration. Big volume and large distribution costs might, for instance, make it advisable to send the user only the updates, as with the paper backup to realtime.

Figure 8-7 compares the classical realtime solution, videotex, and paper handling in terms of implementation difficulty and time intervals for database update. The greatest amount of expert work is with classical realtime. Implementing videotex is very simple.

The VDB may be updated less frequently with videotex than with realtime, but it is far more up-to-date than the information kept in paper records. Besides, managerial solutions do not need realtime update; once every 24 hours is often enough. Accuracy, ease of use, and exception reporting for corrective action are the three issues that are at a premium in management-type implementation.

Access to
the Viewdatabase

*"Almost anything is easier to get into than
out of."* ALLEN'S LAW

Programmers and analysts have been accustomed to consider an application as their own, and that has been true of the files in the database. The screens, formatted to their liking, are theirs. The fact that they should have been sharing such things with the user rarely, if ever, occurs to them. Yet, it is an absolute necessity.

The files and formats of online, interactive systems are the user's own, and users will increasingly address the text and database by themselves. They can get service out of this integration or mess things up, depending on how the system is designed. The user is not working at a technical level; he is simply trying to find out where the text and data are and what they look like. Their characteristics and attributes are the business of the specialist, who must be able to take into full account the user's requirements if he is to perform his task in an able manner.

To meet the user's requirements, the system designer must understand and follow a systematic methodology for the standardization and integration of data resources at the organizational level. The text and database integrate all information elements and interrelate such components across major functional boundaries. A coherent environment must result from the integration of user requirements.

Similar observations can be made in regard to the transactional characteristics when examined from a videotex perspective, except that pages and formats are standardized. All the analyst has to do is assure that text, data, and images are properly stored in the VDB in the first place. The system takes over from then on. It provides (as has been explained) flexible input/output approaches well suited to the user's training level, allows the user to define his set of output views, provides shorter ways to perform user tasks than classical data processing, permits the user to define some particular end use status (or, if the user is a specialist, to make background processes visible), enriches the displayed information by means of encoded messages, provides well-rounded safeguards, assures fault tolerance, gives error messages with correction hints, provides decision aids if tasks cannot be executed as desired, and generally helps the user find his way.

The Access Mechanism

Like any other text and database, the VDB consists of a logical and a physical part. The latter is a well-known memory support such as a disk drive. The former is software-based and has as an objective the management of the *pages* stored in the VDB. A page may contain 24 lines, 40 columns, and hence 960 character positions (including many blanks). Here we are trading simplicity for memory storage. Correctly, videotex uses more storage (the cost of which steadily decreases) rather than increases complexity through compression methods. Furthermore, not all 960 characters in a page contain user information, because some rows are dedicated to identifiers, commands, and *help messages*, as we will see in the coming sections.

Access to the VDB presupposes an infrastructure which must be both internally efficient and easy for the end user to access. That is true of a tree of indexes, each branch of which is terminated by an information page. (In the Philips version, for instance, the tree may consist of up to 99 frames.) A user is allowed access only to a subtree, whose root is not necessarily the root of the complete tree. The remainder of the hierarchical structure is completely invisible to that user. Each user has a unique four-digit passnumber and a four-character password; this makes it possible to give some users certain privileges. For instance, a user who has editing privileges is permitted to modify and extend the database within a given subtree. In extent, it may be either a part of or all of that user's accessible subtree.

In a sense, a user's subtree is his view of the database which he can access. It may define an application or a complete range of applications. Thus, the choice of the subtree's organization and structure involves the first vital steps in using videotex. The user should:

1. Choose the application

2. Define the purpose

3. Identify the basis for selecting data from systems

4. Solve the structural problems regarding the VDB in a way that is sensitive

5. Decide on the pages to be stored on the VDB

Two types of page must be distinguished:

- The *index*, or *routing, page*, which allows the selection of an infopage
- The *infopage,* which contains the data

Index, or routing, pages make access to the VDB user-friendly, but they also give the user two structuring jobs to do.

The structure of the routing tree should follow a logical path, the one we would most likely use in trying to find a page. That of the infopages must meet different criteria. The first basic issue in form design is to decide whether the infopage should be made as (1) a presentation text (as if it were a flipchart) or (2) a decision screen. Both the way of presenting the contents and the fill-in of the page will vary with what is chosen for man-information communication. If it is a presentation text, we should write in alternate lines at the most. If it is a decision screen, we should use a double title, put in headings, leave one line free, and then use every line as in hardcopy printing. In either case, we should simplify what we have; otherwise, it is not readable. And we should also use standard sets whenever possible.

Lack of standardization in data structures leads to interfaces and gateways which, as we have so often stated, result in delays, errors, and costs. Figure 9-1 makes that point by contrasting the simplicity of a single data structure with the complexity of the multiple structures which characterize classical realtime. This can be carried a step further into the internal mechanics of the network, where the simplicity of the videotex service interface is contrasted with that implied by multiple incompatible protocols (Fig. 9-2). Evidently, simplicity has a cost, and it is best expressed by the limitations which are implied.

Within the videotex mechanism are incorporated the message protocol and the user interface programs to assure reliable transport activity. Those features are far more complex in the classical realtime solution, where, as we know, a message is composed of three visible parts: (1) the identification, (2) the command, and (3) the document. Furthermore, with videotex the whole structure is fairly complete. The identification consists of a transaction number assigned by the originating terminal (or host). The command part includes an operation code, an argument list, the destination mailbox, and a list of the nodes and hosts that have handled the message. The document needs a header and a body.

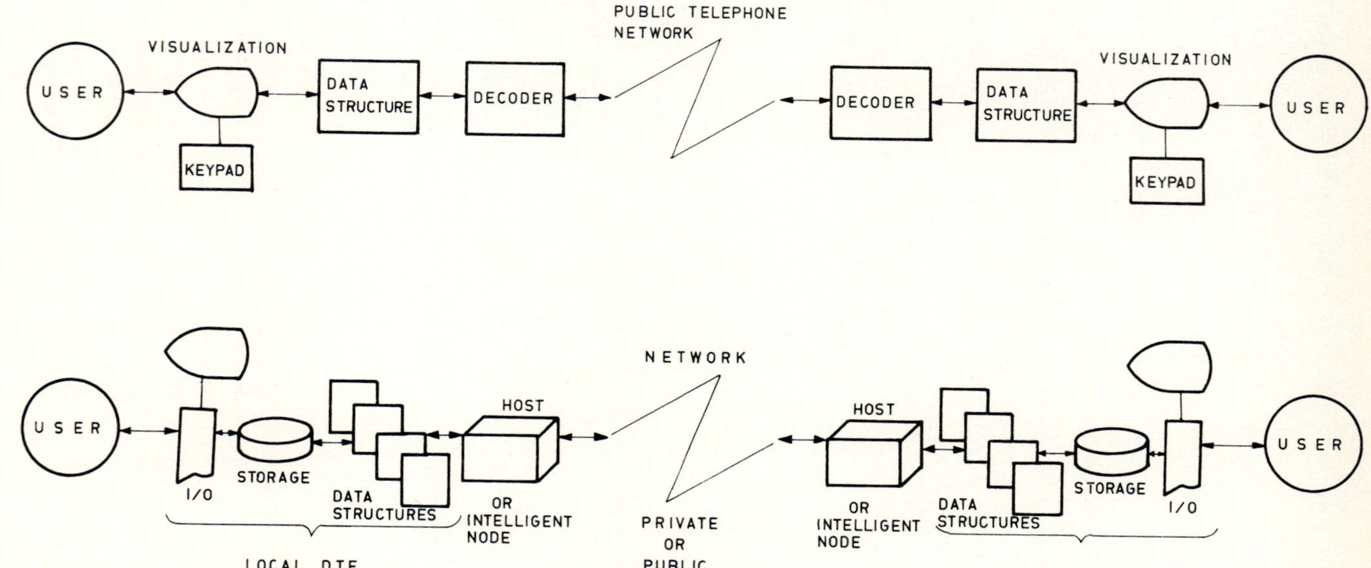

Figure 9-1 The simplicity of a videotex implementation contrasted with the relative complexity of a classical DB/DC network.

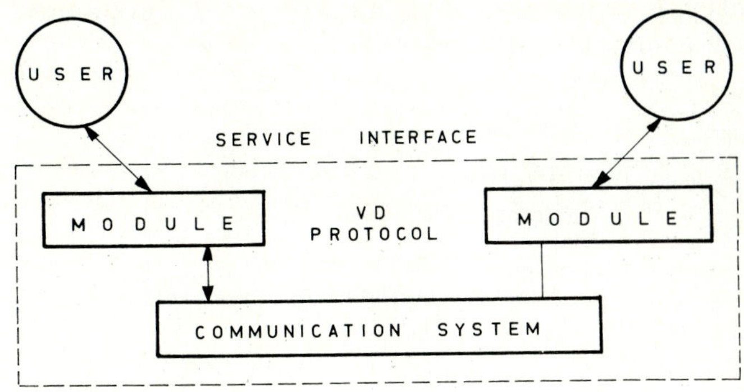

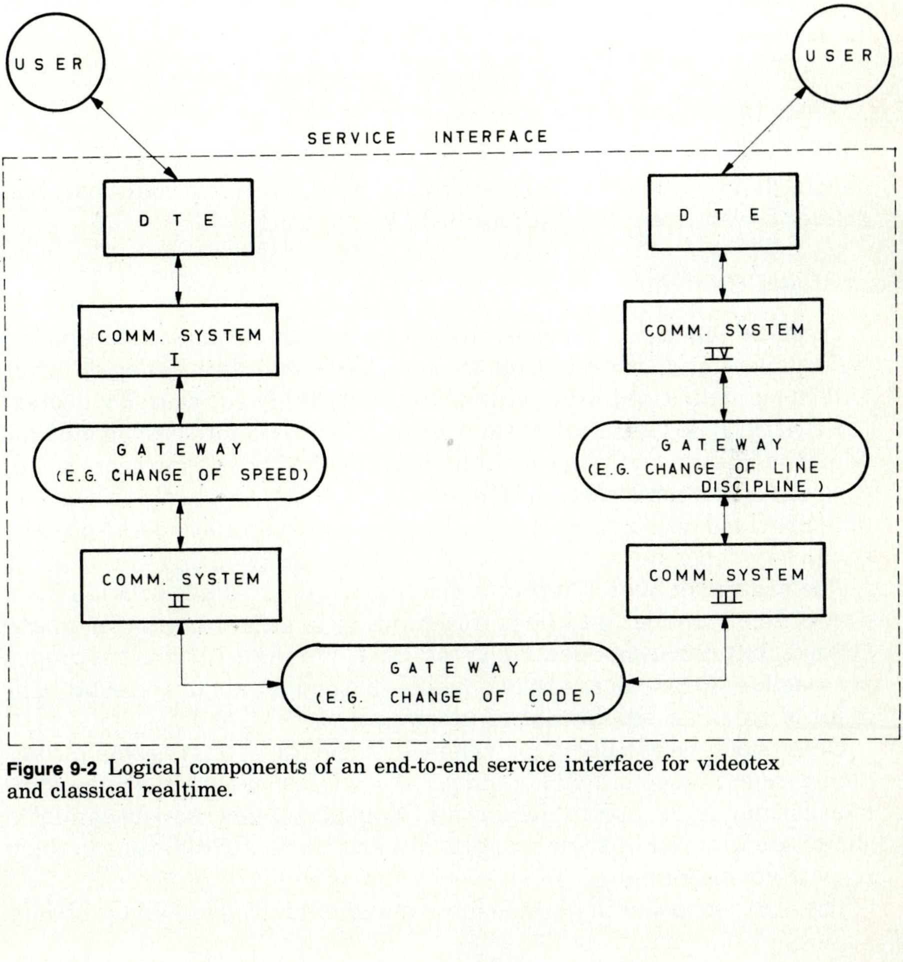

Figure 9-2 Logical components of an end-to-end service interface for videotex and classical realtime.

The following component parts define a message structure. The representation is not necessarily transparent with the classical realtime, but it is transparent with videotex.

Part	*Significance*
Message	ID, command, document
ID	TN, IHN
TN	Transaction number
IHN	Internetwork host numbers
Command	Mailbox, trace, operation, arguments
Mailbox	Destination address
Trace	A list of all the nodes that handled the transmission
Operation	The name of the operation to be performed
Arguments	Parameters needed to carry out an operation
Document	Header, body
Header	The commands specific to document handling
Body	The information carried by the document.

Quite evidently, all commands are overhead. They are necessary to assure delivery of the document to the address in the mailbox—and deliver it errorfree and without delays.

Format and Content

Classical editorial responsibility is for both format and content. Newspapers and broadcasting stations have the duty to decide what will be published and what will be withheld. But in the case of videotex, the British Post Office takes the view that it is no more responsible for what the IP puts on the screen than it is for what telephone users say to each other in a conversation. The IP is responsible for checking that the data provided does not violate the laws against defamation and obscenity or any other laws.

The argument over whether it is appropriate for IPs to provide their own code of conduct and be self-regulating is valid (and so far unanswered), but it does not pertain to this book. On the contrary, the proper perspective is that there are two schools of thought about the establishment of rules for the format of infopages.

One school, particularly prevalent in America, is that, given a common-carrier and open-market videotex service, there should be no common format in regard to aesthetics. Magazines, newspapers, and TV shows are kindred in some respects but are markedly different to their respective audiences.

The other school of thought is more prevalent in Europe and probably has its origin in the fact that videotex there is offered by government monopolies (British Post Office, Bundespost, French PTT, and so on). This school of thought holds that infopage layouts should have a number of common characteristics. Let us look more carefully into what that means.

Frames (pages) should carry headings and subheadings if possible. If for some reason the frames contain scattered information, then a fixed color should be used for repetitive data. When all detailed layout work is standardized:

1. Editing time will be reduced to a minimum.

2. The user will always know where to look for text, data, and image.

Typically, to look for information, the user will access videotex by:

- Index selection (menu)
- Crosslink (alternate routing)
- Page number, directly
- Keyword (selection by subject)

Crosslinks can be established, and the best example of a crosslink is that of a book reader looking at the chapter heading versus the alphabetic index. The principle is straightforward:

- If the infopage at the end is different, then we must reach it by following different routes.
- If the infopage at the end of different routes is the same, then we must cross-index (Fig. 9-3).

Cross-indexing can lead to either an infopage or an index page. Here, again, the rule is straightforward. If to only one page, go to the infopage, if to a whole lot, go to the index page.

A keyword is a word which has been attached to a particular index page of a series of information pages. It is displayed on every page which has one. Supplying a keyword takes us directly from wherever we happen to be to the page to which the keyword is attached. For example, we can get directly to the index page whose keyword is PAYROLL by keying PAYROLL #. When a keyword is attached to a series of information pages, supplying the keyword causes the leading information page to be displayed.

As can be seen in Fig. 9-4 the index pages give the user an easy way to reach the requested infopages. The latter can be a chapter (scroll, multipage) and contain subpages up to the number the videotex software can support. (For instance, the early Philips release supported 19 subpages, which has now been upgraded to 99.) This multipage (scroll) concept allows more than 24 lines to be handled within the same frame, which can run, for instance, up to 99 subpages. [Of the 24 lines, the first two should be kept for identifiers (keywords, dates, etc.) and the last two for help and go-to messages.]

Even if a system does not support direct access to subpages, the latter can be reached by browsing. Hence, in general, access can be direct and in a serial manner. This will usually be for less important information or continuation data. Another requirement is addressing the *complement* of the page we are looking at. Such is the case with additional airline routings needed to reach a destination.

Going back to the definitions we have given, it will be recalled that a frame (subpage) is a screenful of information. The page itself (more precisely, the multipage), is a *node* of the DB. Hence, it is an addressable unit. What makes the definition process so important is that access is governed by indexing, and menu selection calls for preestablished

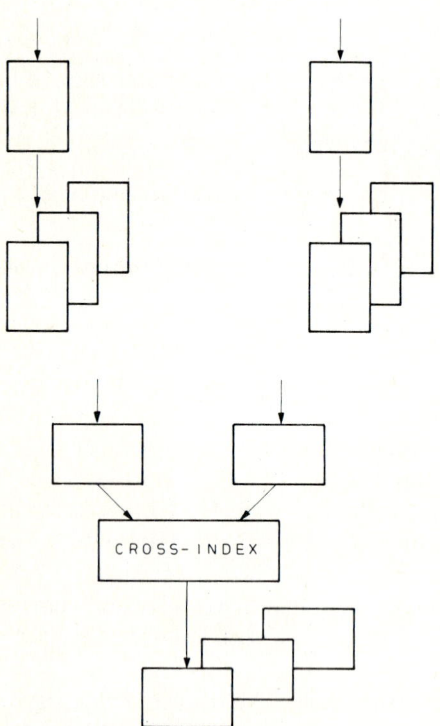

Figure 9-3 Indexing and cross-indexing facilities are fundamental in videotex implementation.

routes. Figure 9-5 gives three examples of alternative selection routes. We must establish which one will be the most likely, and to do so we are obliged to project and predefine the routes people will go through looking for information.

Routing (particularly flexible routing) can be page-consuming. In many videotex systems, indexing uses more pages than the information, but usually the ratio of indexing (routing) pages and infopages is 50-50. So if we need 1000 pages for infopages, we must expect 2000 pages in memory and never fail to recall that the more the indexing pages the greater the system flexibility.

Menus

Though fundamental research in the design of menu networks is still lacking, some basic principles have evolved from the analysis, construction, and maintenance of such systems:

1. Provide from the beginning a structured approach, and establish a monitoring tool which helps record the frequency of use of the different page search paths.

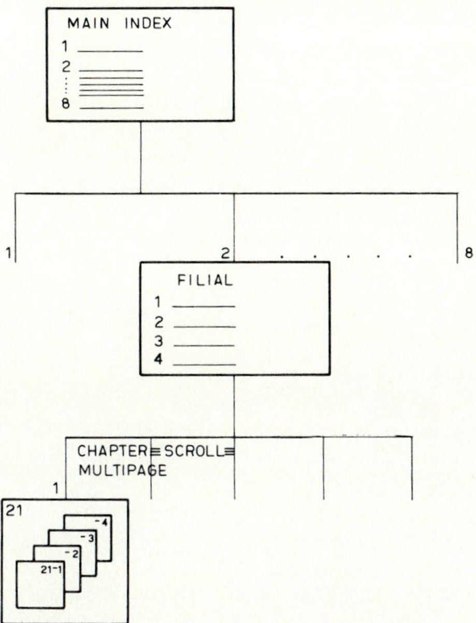

Figure 9-4 Main indexes and filials provide an easy way to reach the requested infopages.

2. Based on operational statistics, identify the most highly traveled paths and establish direct routes (shortcuts). Keywords and go-to statements can be helpful. Quite similarly, be ready to remove direct routes which are not frequently used but complicate the overall structure of the menu network or violate the logical data flow procedures.

3. If your research (or operational statistics) demonstrates that there is a stable, known, and properly understood underlying information structure, allow the menu system to reflect it. That is what is meant by a *natural data flow*.

4. If, on the contrary, natural data flow does not exist, then adopt a top-down structured design.

Basically, a structured design will involve principles known from networks; some basic examples are hierarchy, concatenation, if-then (op-

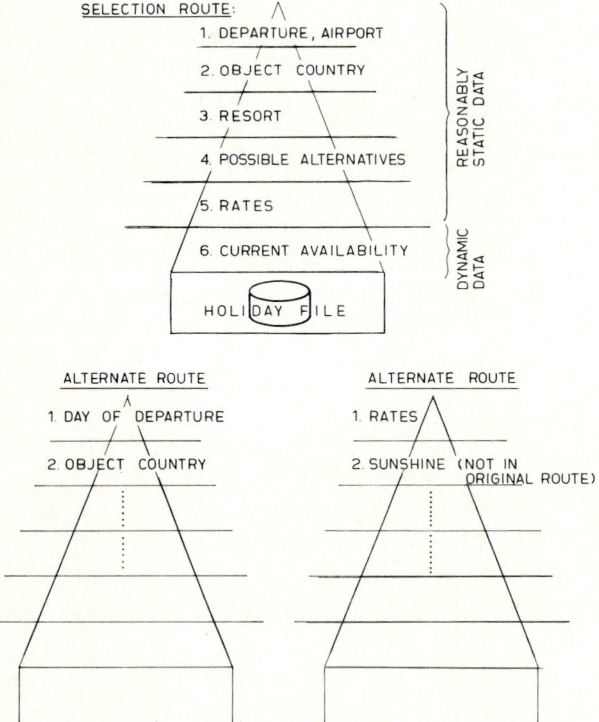

Figure 9-5 Alternate routing is possible, and each route features its own selection criteria.

tion), if-then-else (selection), and iteration (loop). We will briefly review their use and impact.

The hierarchical structure is a cornerstone in menu selection the goal of which is to present the user with a logical sequence of routing pages and infopages (frames). This orderly presentation should support options and allow the user to choose where to go next from a set of properly defined possibilities. The selection of an option on a routing page represents the transition from one node of a network to the next (logical graph). The routing pages contain the directions necessary to link themselves to other routing pages or to infopages in the VDB. In a given videotex system the routing pages can number in the hundreds and help to manage thousands of infopages.

Within a given hierarchical structure, *concatenation,* or sequence, is the simplest interconnection form. Combined with *iteration,* it makes it feasible to proceed to the next node and then return to the node of departure. Iteration is valuable for another reason also. A *help command* is its simplest form. Typically, a help command branches into another frame for identification or explanation purposes and returns to its frame of departure.

An iteration node may, however, also present *selection capabilities,* which give the user the opportunity to make decisions on where to go next. An example is the if-then option often used as the entry point to a menu system. Help frames, for instance, may provide an explanation of how to use the system and proceed to the mainline without the need for return to the original frame or for user action.

If-then-else is a more complex form that gives the user a choice between two options. In its pure presentation, it is a restrictive aspect of branching: Usually selection keys offer up to 10 branch capabilities and at times much more than that. They are achieved by combining selection nodes in sequence quite similarly to a step-by-step telephone branching system. The basic nodes referred to are fundamental parts of a menu network. Composite structures that can be built from them permit large menu systems. Such networks will be characterized by nesting and concatenation. In the more complex menu network, concatenation connects the outgoing link of one substructure with the incoming link of another. Nesting replaces a single exit node with one which presents one forward link and one return (to the origin) connector.

Since menu structures tend to become complex and involve a number of intermediate nodes, the selection of the object infopage may, at times, be a long and cumbersome process. It calls for a direct connection bypassing the intermediary steps—something like Fortran's GOTO statement. Direct addressing can be done by such a means or with keywords, and it permits the user to arrive instantly at any wanted infopage.

Routing

Videotex, it was stated, supports two types of pages: index or routing pages (taxonomical, routing frames) and infopages or end frames (pages, subpages, infopages). They can be accessed through menu selection or directly (numeric or mnemonic). In the latter case, they must be numbered, and it is advisable not to mix the numbering systems of the end frames and the routing pages.

The routing pages are best served through a classification code with taxonomical characteristics ideally organized in a relational manner—family, group, and class—down to the level of desired detail; family is the top classification. It is advisable, when the families are built, not to use more than 5 out of 10 choices per digit employed and thereby leave room for further expansion. Transparent to the user is a 10-way control block attached to each routing (index) frame. It includes the security tag which assures authentication (Fig. 9-6), and it is used for selection within the tree structure.

One way to look at the family code is as a two-digit number. The two digits have the benefit of a matrix presentation and, therefore, easier visualization for planning purposes. If this is followed through group and class, we can easily have successive levels of exploration in the menu, each using availble choices for data presentation. Of course, nothing prevents employing more layers if the reporting structure calls for them.

A special problem presents itself when we convert traditional paper-bound reports to videotex. A computer printout typically has 132 columns (we do not always need all of them, but we use them), and it can be run on one or more sheets of ply paper. Even if only one sheet is used in the report, that will mean $132 \times 60 = 7920$ positions or roughly 10 infopages of interactive videoform. (The relation is not linear, because headings and identifiers need repeating, but neither is it advisable to

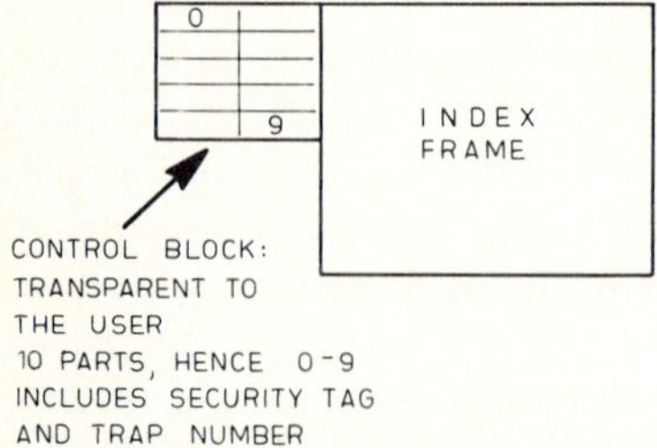

Figure 9-6 Block control applied with an index frame provides authentication and authorization.

convert hardcopy to softcopy one to one. They are totally different concepts. The preceding material is offered only as an example.)

To head a group of related pages (usually, in this case, called subpages), there is the title page (multipage). It fits somehow between the image (and the function) of end frames and that of routing frames. The title page *is* an end frame, but it serves no information purpose; it is another sort of routing.

Let us first see how to identify the infopages (end frames). The most advisable approach is to use a *running number* absolutely independent of the classification code followed with the index pages. Do not try to do both with one code; the effort will fail. Infopages should get numbers of ascending series as they come into existence, which can be called

<basic code>   or   <bc>

That is an identification (ID) number which itself bears no information as to the infopage's content and purpose, which is the object of the tree structure.

Being an end frame, the multipage will be identified through a <basic code>. Like any other <basic code>, it can be numeric or mnemonic, although in the general case, because of the numeric keypads, numeric should be favored. But the title page, we said, is another index. Behind it may come several infopages (subpages). How are they going to be identified? The answer is through a <suffix> to the <basic code>.

Prestel uses numeric codes for the end pages and adds one letter of the alphabet for the suffix, which limits <suffix> to a maximum of 24 possibilities. Other systems, Philips for example, use a two-decimal-digit <suffix> and so make available up to 99 subpages for each title page.

Title pages are not supported by all systems. When they are, the content of the subpages attached to them can be looked up through <suffix> access or browsing. (This type of access is one of the key differences between current videotex systems.)

Two-Key Choosing

We said that a routing page can have up to ten alternative path possibilities but no more than five should be used in the first instance to make future expansion feasible. There is a way to bypass this limitation, provided the videotex software supports it. It is known as *two-key* choosing. It makes putting more than 10 choices on the same page possible, and it may lead to browsing by hitting the "next frame" key on the pad. [The two-key choosing (TKC) capability is more important if we can manage to have, say, 20, 30, or 40 choices instead of up to 10. A

TKC extension called numeric key choosing (NKC) permits handling a four-digit choice by accessing this routing page through XXXX.]

Another solution is the mnemonic (alphabetic, alphanumeric) keyword. Still another is *logical mapping*, direct access to infopages by making the page number a direct reflection of the information it contains. And there is the *escape option*. The intent is to allow the user to back out (if an erroneous choice was made), close down the DB contact (if satisfied), or return to top of the tree to make a new search against another criterion.

Let us take one more look into the use of the suffix to identify information subpages. All frames forming one series are numbered in sequence. Upon reaching the desired infopage, numbered through <bc>, we first obtain a *contents page*. The information subpages are numbered by <s> from 2 onward up to 99. To move from any information page to the following one, all we need do is press the go key (#). In that way, we may read straight through a series of infopages if we wish. If we press the # key when on the last page of the series, we will be taken back to the last index page again.

Since the route structure is orderly (and that is what makes menu selection so friendly to the end user), it should be evident that it is easy to create the right VDB from the beginning. But it is costly and confusing to do it afterward. To lay the foundations for the right VDB in the first place, we should start by looking at what the user wants as information:

1. Define (and write down) what we wish to get from the system for each user who will have access to it.

2. Decide on authorizations, updates, timing, ideal number of pages, and maximum number of pages.

3. Identify how the users will access the information.

And, as we saw, it is good practice to divide identification from classification. The latter, being taxonomical in nature, should be the routing structure to the infopages. The infopages (multipages) have to be structured, so that whoever goes into that data gets the same information. Furthermore, experience suggests that data is mobile. And we must allow for expansion.

The VDB structure should reflect (and be controlled by) the way the information will be retrieved. It cannot be repeated too often that we must predict how the user will call up the information. With realtime we can select from different fields in the database; with videotex we must preestablish the connections. There are many combinations depending on the data we need, but we must establish the relations between stored information elements (IE), frames (pages), and pages (multipage). Browsing routines are used to go to the next multipage. In that sense, key choices may not only go downward but also go upward and sidewise.

Response frames support two-way communication. And with most videotex systems *confirmation pages* are talked about quite a lot as a coming facility. If we look for an index (routing) page, we must key in a selected page number; if for an infopage, we should key in the next page number or a new page number; for a multipage, handle videotex like a chapter in a book—browse through.

A Good Editing System

A good editing system is fundamental. It presupposes an online facility permitting the manual entry to, amendment of, deletion of, overwriting of, and copying database frames. The process carries out user password and page ownership checks and accepts control information related to each frame such as routing detail and closed user group particulars.

Another basic facility to be supported by the videotex software is bulk updating. The reference is to an editing process which allows for the retrieval, insertion, replacement, and deletion of information frames whether in batch mode using magnetic tape as the input medium or in realtime via computer-to-computer communications.

A third essential is a user response facility (URF) to enable terminal users to send messages to IPs who have set up response frames within their databases. The user's name, address, and telephone number can be automatically associated with a message which is then stored on disk for subsequent retrieval by the response frame owner.

We have also spoken of closed user groups. The videotex software system offers two different types of such facilities. One gives the IP the ability to restrict access to some or all of his data frames to specified terminal users. The other allows the IP to temporarily block all user access to his database. With videotex, as with all online systems, log-on and log-off are fundamental requirements. Log-on validates the terminal identification and pass number of each user before access to the videotex system is permitted. Log-off allows a user's call to be terminated in an orderly fashion and updates billing information held on the system user file.

For a public or corporate videotex service, user billing also is basic. A facility should provide information to both the user and the videotex system operator. A bill frame, accessed via the normal user procedures, displays the accumulated cost of frames retrieved during the user's current session, the total value of the user's account since the last bill was sent, and the date and amount of the last bill.

For IPs a revenue table must be maintained. It must be continuously updated to show the revenue attributable to each IP for retrieval by users of his frames. The table may be printed out at regular intervals to facilitate the manual preparation of IP accounts.

A trap number facility is essential for direct retrieval while editing. The trap number of an index entry is always identical with that of the record to which it points. (As indicated in the preceding section in the discussion of the control block, trap numbers are held in index entries to determine if the record pointed at is accessible to the user while the index page is displayed). Unless the editing user has the trap-setting privilege, a record when created always has the same trap number as the index to which it is connected. An editing user who has the trap-setting privilege sets the trap number of an index entry when he creates it. Such rules are sufficient for the creation of separate tree-structured databases within the one tree (Fig. 9-7).

Security of access is assured by such features as:

1. User number and password to get into the VDB

2. Passwords held internally in an encoded form

3. A trap system to control accessibility of data records

4. A privilege system to assure the use of the editor and system functions

Users with any privilege (use of editing or system functions) must go through a two-stage log-in process. Provision is made for monitoring

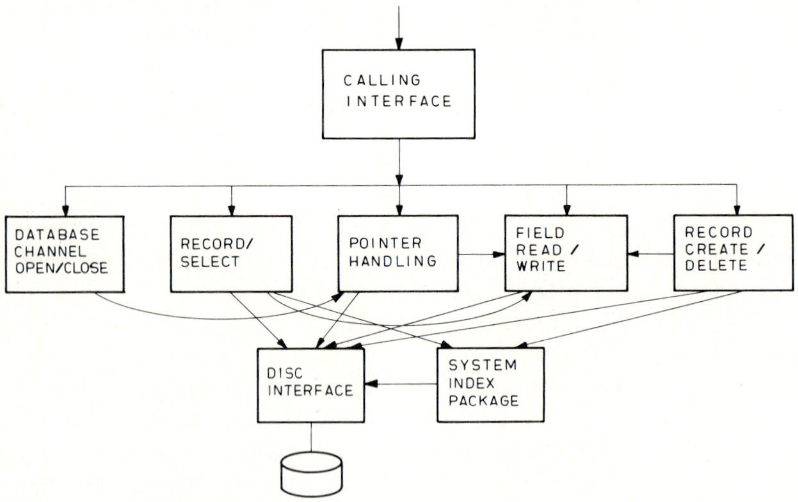

Figure 9-7 Separate tree-structured databases within one master tree and integrated through a calling interface.

the opening and closing of tasks, logging the users, and error occurrences. (The details of task number, user number, and error number are recorded.) The total number of indexes and information pages accessed and the functions employed by the user are recorded on log-out. Access of individual pages is not recorded.

An error recovery mechanism is provided. When an error occurs while a user is logged in, the database channel is closed and reopened, and the user is returned to the main index. An error message is displayed on the bottom line if a user input cannot be interpreted. Pressing the backspace key (←) cancels the immediately preceding character. Any number of characters may be canceled by pressing the backspace key the appropriate number of times. In some circumstances, the character * may be used for correcting errors.

10

Editing

*"Comedy is life looked at from a distance.
Tragedy is when you take a close look."*
CHARLIE CHAPLIN

Editing is one of the fundamental processes in videotex, and it concerns everybody. A user with an editing terminal can issue videotex pages, and his work resembles that of the editor: preparing for publication (on videoscreens) by selection, arrangement, and annotation. The editor sets the layout, chooses the colors (background and foreground), selects the upper- and lowercase letters, and builds the graphs. He has at his disposition a videotex character set (Fig. 10-1). He also follows the menu selection route in designing the index pages.

- Figure 10-2*a* shows a routing frame for visiting New York designed by *The New York Times* as an IP. It presents the decimal key choices (up to 10 selections) we have outlined.

- Figure 10-2*b*, a Prestel index frame for mailbox, illustrates the two-key choosing capability referred to earlier.

Figure 10-1 A videotex character set.

The editor has the responsibility to safeguard and enhance the user's ability to do productive work; videotex use should present the greatest convenience to the person with the problem. The aggregate user time is more costly than the computer time which is involved. Another basic consideration in projecting interactive systems is that not only will some of the authorized users work with the system at the same moment but also a small fraction of them will account for the largest share of available resources.

Statistics gathered from interactive systems indicate that less than 10 percent of the users consume about three-fourths of the computing resources. Such an imbalance characterizes the distribution of computing usage in many installations, and it should be kept in mind in system design. That is another way to say that the end user should always be brought into the picture. Experience with conversion from batch to interactive systems demonstrates that the user is also happy to take that step.

Overcoming procedural handicaps should be one of the primary goals on the new frontier of system design. Handling psychological barriers requires new ideas. To overcome fear of the system, a leading company let each manager learn how to run a simple conversational program. All the managers became excited at the system's possibilities, whereas they had not quite understood the printouts they had been getting in the past. With videotex we are in a fine position to do a clean editing job from the beginning. Let us use the opportunity.

Index and Page Editors

Two editing systems, one for index and one for page, are available to those with the appropriate privilege. The *index editor* provides for the insertion, deletion, or modification of entries in the index the user is

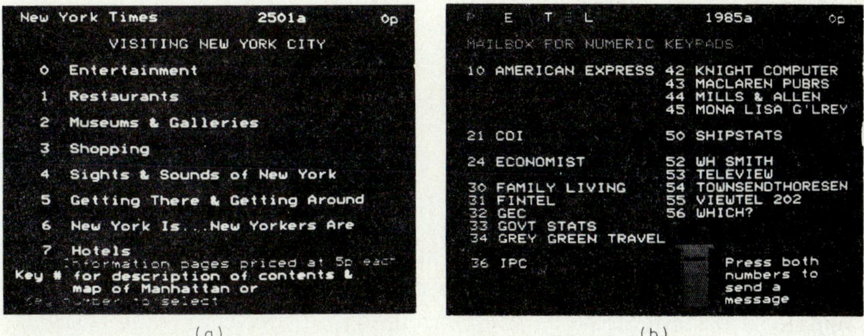

(a)   (b)

Figure 10-2 (*a*) Itinerary for visiting New York designed by *The New York Times*. (*b*) A Prestel index frame for mailbox.

currently viewing. It operates by displaying the electronic equivalent of a form and guides the user through the process of filling it in. As entries are made on the form, checks for possible errors are performed. For example, deletion of an index which is not empty or to which a cross-reference exists is prohibited. When the form has been completed, the required modifications of the VDB are automatically reflected in the contents of the corresponding index page. There is no possibility the choices offered by an index page will not correspond with the tree structure.

The index editor provides the following facilities:

1. Insert entry
2. Change entry
3. Delete entry
4. External file
5. Edit upper text
6. Edit lower text
7. Quit index editor

Function line traps provide for read and write security and allow access or write authorization.

The *page editor* makes it possible to prepare and modify individual frames. As well as allowing the straightforward modification of the contents of a frame by overwriting, it provides a number of text handling facilities including reformatting after the insertion or deletion of text and repositioning text on the page.

A basic videotex characteristic is the organization of the user segments, which have six components:

- Frame control area
- Frame
- Frame table
- User table
- Input buffer
- User work area

The frame control area contains the user number currently connected to the port number corresponding to the segment and the sequence of four characters made up from one form feed and three carriage returns.

The frame component consists of 960 bytes of storage and receives from disk the data to be displayed to the user. It is not always full because, on the average, the content of a page displayed to the user is only 500 to 600 bytes long. A substantial part of it must be left unused to meet editing requirements of appearance, tabulation, and so on.

However, since the maximum number of character positions in a displayed page is 960 (24 rows of 40 characters), the full amount is allowed in the frame region. The contents of the frame region may be modified, in certain cases, before being sent down the line to the user. Accepting that, on the average, the page content is about 500 characters, the average cycle time for the average user is roughly 20 s (seconds). A videotex computer using a single moving-head disk would be able to service users within the stipulated maximum response time of 2 s.

The frame table is 64 bytes long; it is compiled at the same time as the frame, and it reflects the data structure of the information provided. It is stored with the frame content in disk storage and transferred to main storage at the same time as the associated frame. The structure of the frame table is shown in Fig. 10-3. The content of the frame table determines the action of the system following the user's response to a displayed frame. The frame table contains the:

- Identity (such as number) of the page and frame

- Task number—for instance, the process number which deals with a user response related to the frame

- Actual disk addresses of the next frame and the frames corresponding to the choices offered in the current frame

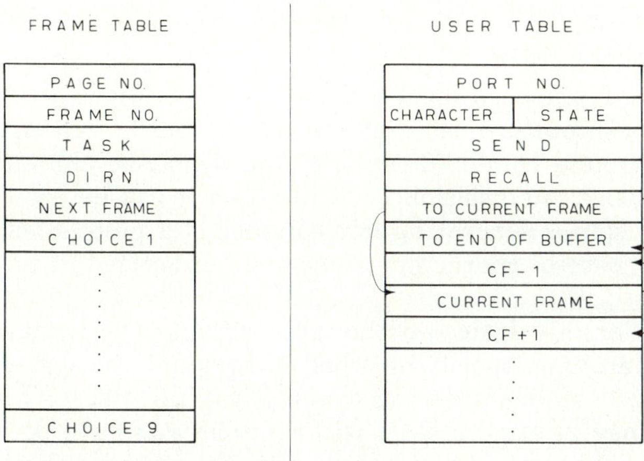

Figure 10-3 The structure of a frame table and of a user table. The latter is addressed to the user transactions.

It also includes direction (DIRN), which is used by the task number to decide whether the frame is to be output directly (DIRN=0) or sent to another task specified for modification (DIRN≠0).

Correspondingly, the user table contains details pertinent to the user's current transactions. Examples are a buffer which maintains a record of the current frame (as a disk address) and the address of the previous choices, which enables a user to retrace his steps. Frame addresses are entered in the user table by the DISK process and retrieved when necessary by one or more of the TASK processes. The user table also contains PORTNUMBER, which identifies the computer port to which the user is connected.

The character set supported by videotex assures:

1. Continuous graphics
2. Separated graphics
3. Color
4. Flash
5. Background
6. Double height
7. Alphanumeric data
8. Lowercase and uppercase characters

Color for background and characters and double-height characters is obtained by keying in information as to:

- Background color
- New background
- Double height
- Color writing

Each operation needs one position. Thus if all four operations are to be performed, four positions are necessary. Two positions are enough for change in background color. For each of the double-height and color-writing operations, one position is plenty, and it can be easily hidden in the blank spaces separating two columns of a table. Flash will require another control character, which can be inserted at any point in the line.

None of these positions are dedicated; they are available if we want to use them. If, for instance, we specify no other background, the system will start with black. In other words, the system says: "Till you tell me something else (change in attributes), I will continue as I have been basically instructed to or as you have told me with the preceding command." (There is an alternative to this approach. The French system divides attributes from text by using 2 bytes/character, one for each. But that doubles the memory requirements.)

The Page Mechanism

Recommendations can be made in regard to the esthetics of page layout. They concern the use of color, standard frames, the use of lines, and readability of print:

1. Do not use more than two colors. Three are acceptable, and four make reading difficult.

2. Use colors which are close together in the spectrum. (Colors excite different areas of the eye, and the far out do not allow good focus.)

3. Try colors with similar luminance, unless you wish to highlight.

4. Use background color to build a "house style," but use it with extreme caution. Otherwise, it is hard to read.

5. In color graphics, the colors up the spectrum (white, yellow) tend to move on the eye. Dark colors stay better.

6. In developing a house style, avoid empty lines, particularly at the top. Similarly, put graphics at the bottom and not at the top of the page.

Though it is a fact that much of page design is determined by the facilities supported by the different videotex releases and standardization has not yet taken place, there are enough similarities for us to use a practical example to follow the editing procedure.

Let us suppose that Fig. 10-4 presents the layout of the editing unit;

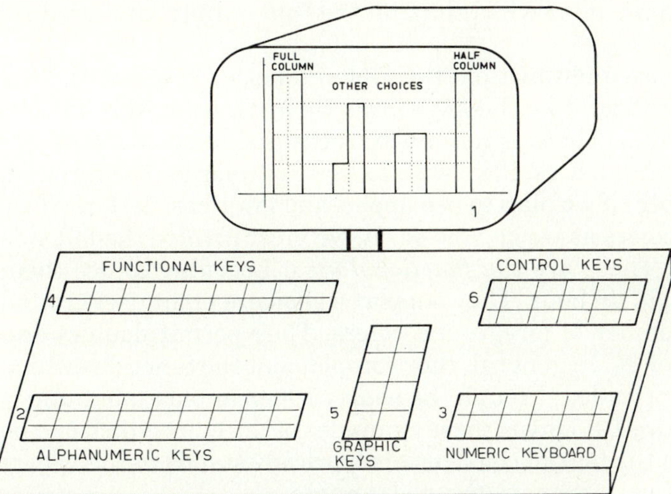

Figure 10-4 The editing terminal used in the experiments with a graphic display on the videoscreen, and a range of functional, numeric, alphanumeric, graphic, and control keys.

give or take some variations, editing boards have a general structure. Six main groups can be distinguished. Together they form the editing terminal, which can be used for both routing frames and infopages:

1. Color video unit
2. Alphanumeric keyboard
3. Numeric keyboard
4. Functional keys
5. Graphic keys
6. Control keys

The *color video unit* is typically a TV set with 24 by 40 positions and color and graphics capabilities. The information which will be written on it, a combination of text, data, and image, will be stored in memory (through the appropriate routine) and presented to authorized videotex users. It is, therefore, of the utmost importance that the information be correct, and the immediate visual feedback afforded by the video of the editing unit is most helpful in this respect.

On the editing terminal the infopage format will be prepared and the content filled in. Format is established *once* for each frame and used repetitively thereafter with variable text and data unless there is a reason to change it. The data can be inserted through the editor but the work is not exciting; it is a routine procedure which should be automated. In fact, as we will see in the following section, there are editing programs for mainframes which handle the "old output" and produce videotex frames.

Both in structure and function, the alphanumeric keyboard and the numeric keyboard look like the keyboards we know from word processors and calculators. There is not much, therefore, to be added except the fact of the prerequisites. If we wish to write in color for background, foreground, or both, if we plan to use upper- and lowercase letters, if we want to change colors as we go, and so on, we must involve the left side of the keyboard. These are the *functional keys*. Learning to use them takes less than half an hour. They support a choice of eight background and an equal number of foreground colors. They permit double- and single-height choices. In general, they complement the functions which were classically provided through the nearly 100-year old alphanumeric keyboard. Figure 10-5 gives three examples of the edits which can be obtained, though the lack of color in this hardcopy stands in the way of appreciating the variations in the background.

The *graphic keys* represent one of the strokes of genius in videotex. The use of these keys is child's play: It suffices to choose the color one wants (in the functional keys), depress the graphic keys to show on video the graph which is to be made, and finally, through the repeat key of the keyboard, imprint that graph on the softcopy.

Note that each of the 960 characters we can write on the face of the video for each and every infopage can be split into parts. The British system supports six parts; the American (made for GTE) supports eight. (An example is given on the video of Fig. 10-4.) This means that we can reach good enough precision in graphics at an eighth of a byte, which is not, however, a bit level. Histograms are perfectly drawn this way, but circles and pie charts are not. That is why some people call the method *semigraphics*.

Graphics can also be created with a light pen as an input device. A more complex approach is to use a graphics camera unit, which can digitize a picture by using a monochrome camera. Neither solution is advisable for a beginner. If there is good advice to give, it is *keep it simple*.

Figure 10-6 exemplifies what we have been talking about. There is a histogram (in color in the softcopy), a publisher's house emblem, a lighthouse, and Peanuts. As can be seen, the histogram is perfectly

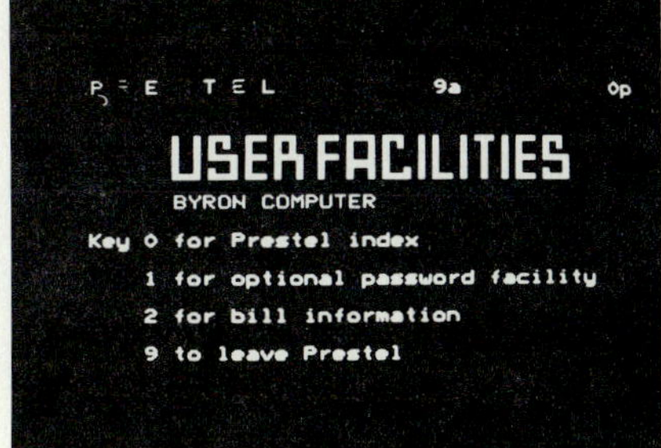

Figure 10-5 Examples of the edits which can be obtained through an editing terminal. They are stored in the viewdatabase and can be retrieved at will.

drawn, the house emblem presents minor problems only in the lower curvilinear parts, the lighthouse itself looks good but the light beams and sailboat's rectilinear design are less than perfect, and Peanuts on softcopy offers nothing less than the hardcopy alternative.

The *control keys,* like the other four sets of keys, can be learned on the first or second try. A few of them, such as spacing, are like those of the classical keyboard, but generally they are oriented toward the handling of the editing functions: move up or down or shift left or right to position the cursor, and so on. These are the elements of the editing terminal and any person with average IQ can easily learn them.

Entering Data into the System

One of the key challenges is how data can be entered into the videotex system without any manual procedures. Since, by being user-friendly, videotex constitutes an ideal solution to decision support, management information can be extracted from bulk processing (batch or realtime)

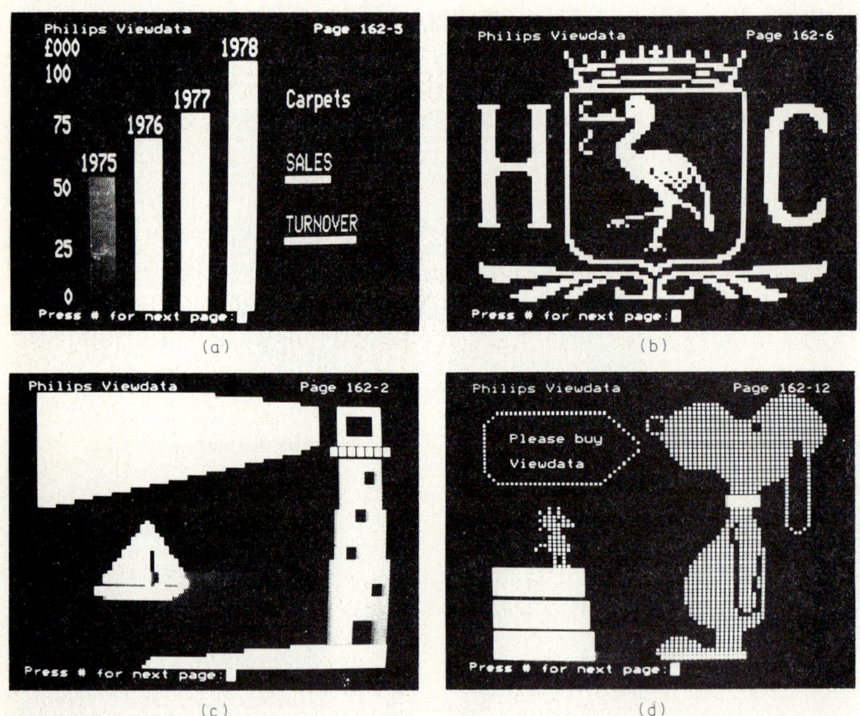

Figure 10-6 Alphanumeric graphics handled by Prestel: (*a*) a histogram, (*b*) a publisher's house emblem, (*c*) a lighthouse, and (*d*) Peanuts. (Observe the perfection of histograms and the difficulty of producing diagonal lines.)

through converter routines able to pick up for videotex presentation information originating elsewhere. This should enable existing computer files to be automatically structured and formatted for loading into the videotex system. Prerequisites have to be met. The VDB must be specially structured to allow the average user to route himself to the required information by making simple choices from a number of offerings at various levels of index.

The system should support a fully structured database incorporating all data access routes, cross-references, and other indexing aids. The structure and associated linkages must be kept up to date, and these structuring tasks must be handled automatically. Hence, the conversion software must contain facilities for generating multiple, multilevel access routes automatically while compiling and maintaining complete database references including all linkages in videotex format. Furthermore, changes in database structure or format should be easily introduced if need be, together with machine-readable information edited automatically to abridge, expand, or supplement coded or textual data.

Facilities must be provided for the automatic formatting of information pages with full control over color, case, flashing, and position in any field. Standard page headings and footings must be specified. A primary index to each end page should always be generated, and alternative indexes should be created as required.

For mapping to the VDB, the IP should be able to specify entry page and start pages of alternative indexes and any action frames. The conversion package should automatically fill all pages from the specified entry points downward without leaving blanks and convert the whole structure to the correct code and format for bulk input to videotex. It should allow parametric approaches to the likely access routes to information. That is necessary to supply a fully structured VDB incorporating all access routes, cross-references, and other routing aids and keep the structure and associated linkages up to date.

Furthermore, information from existing systems may be supplemented to produce formats suitable for publication. Thus, coded information may be expanded or replaced by full text descriptions. Fields may be edited, and uppercase text may be converted to upper- and lowercase. User entry points are provided for special requirements. Routing messages, prompts, and references to response frames may be specified for automatic incorporation into pages. Routing to the next page or the preceding page may be specified, and that provides a basic browsing facility. All linkages associated with routing messages are generated automatically. Alternative indexes to the same information may resequence the same fields or replace some or all of them.

The ideal converter system would be able to read any file organization originated on any make or model of a computer. Unfortunately, the ideal is not yet available. A lesser demand is for the capability to modify a record and make each record in the input file accessible to the user. The user may add, delete, or modify records and also force end of file, limit input file size (useful during application testing), add, delete, or create new fields, concatenate index term and comment for data in the lowest-level index, add run time data, and concatenate textual heading with data so that both may be conditional on occurrence.

The main purpose of the conversion routine is to transmit updating information. As with all data and text handling systems, characters go into memory to create a page of information. And in a coded form, data and image are interchangeable.

Yet, unlike classical data processing, videotex presents a series of user-oriented requirements. For any point we must define two pieces of information: how bright and what color? That is not the usual thing with DP. Paging and presentation perspectives must be protected either automatically through the conversion routine or by editing. To make reading easy, we should not fill up the page. *We have to think of reading, not of writing.* Formatting has to make reading easy. To plan correctly we must dissociate the format from the content, we have often emphasized, and then take the data from the computer and edit it in the selected format.

It is also important that, although the routine conversion work is done automatically, the user be given the capability to generate logical index terms from the actual index, create display fields, use flash or color for exceptions, and modify fields. The latter may include extended field editing, editing of variable-length fields, color and flash by data content to allow for specific words to be highlighted, and so on. Other modifications may involve invoking double-height characters, adding background color, changing index frame layouts by level so that they are specific to the index terms being displayed, enhancing index frames with graphics, converting data frames into response frames, routing from data frames to page numbers contained within the data, deleting the main index tree, and deleting the upper levels of an index when manually entered frames might be preferred.

Mapping requirements also are present; examples are mapping info-pages to user-selected page numbers, and data frames to page numbers contained within the data. It should also be possible to adjust index frames to support logical key choosing and skip past indexing sublevels to browse index terms.

Equipment characteristics should evidently be reflected in the conversion routines. For instance, dot system writing reflects the construction of television sets; hence the 40-column limitation. (Furthermore, if 80 columns were used, it would take double the time to transmit a page.)

File-Handling Procedures

Figure 10-7 demonstrates a file-handling procedure for files A and B, the creation of a common file AB, and the projection A' and B', respectively, related to the basic files. The merging of files AB is the function of a merge program; that of converting the file into VDB is a function of the specialized conversion routine. There are two ways to look at this job.

1. A *full* publication system: processing the master file and creating the VDB.

2. A *partial* publication approach. Instead of processing the master file, we process the transaction file.

The latter will work only on the pages affected by updating, but it may also handle pages which "click" together. For instance, "partial" may be the audit trail. A third solution now in development is to handle the formatting of VDB in realtime. The first known application is in the London Stock Exchange.

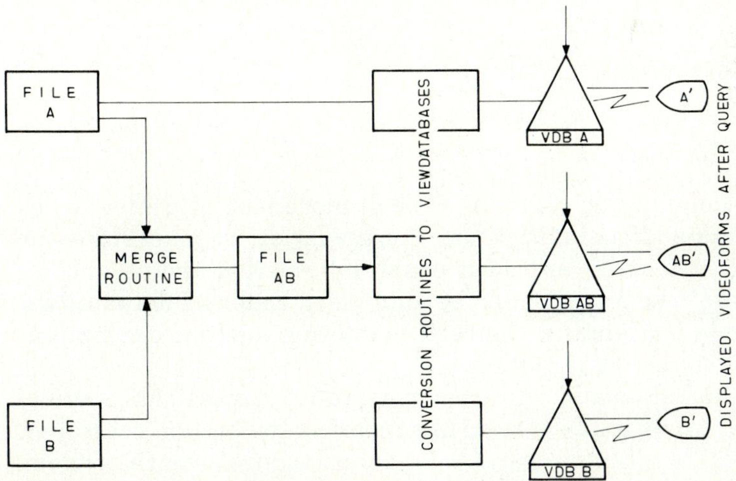

Figure 10-7 File-handling procedures in a viewdatabase and the creation of a common file AB.

A real enough time solution used by airlines called on three codes on Prestel:

- Seats closed as of last night
- Good availability
- Low availability

The minimum update cycle for real enough time is daily or weekly, but it really is related to updating frequency. In this capacity, Videotex might serve as a communications front end to the main VDB. In the travel business, for instance, some 70 percent of calls to an agent do not result in booking, but they keep the operator busy. This concept is similar to cash distributors and automated teller machines. Through it, videotex can be used as a filter to provide RET information. Realtime can then be applied to actual bookings. It is also possible to implement a gateway to make the videotex unit transparent for the 30 percent who decide to book.

Let us review the sense of what has been said in this chapter. We need an automatic system to update the VDB, and this program should be driven by parameters. But for any computer-run system to be effective, we must define the rules for VDB formatting.

1. *Rules* concerned with *information display* (report writer)

2. *Rules* concerned with *updating*

Computer-to-computer transfer is not a problem, but formatting may well be. We need to convert information:

- Format it.
- Make it more understandable.
- Add color.
- Restrict it to 40 by 24.

This can be done in two ways: in a batch mode (one at a time) or by separating format from information. For the latter, we must store—on the videotex computer—both format and information and provide for regular update. The process must be automatic. Entering data into the videotex is easy but lengthy. Converting information from a computer-run database is the best solution.

But even an automatic approach may call for occasional manual intervention. An example of needed support is the system controller, which enables the videotex operator to exercise control over the system. At system start-up, this facility:

- Carries out disk consistency checks

- Reports on any missing or duplicated information nodes in the database

- Sets up system tables

For normal running it holds the system operational parameters, such as the limit of permissible simultaneous users, and originates all line-24 messages. For system shutdown it ensures that a correct and orderly procedure is followed.

Another facility is the so-called videotex archive. This is a process by which the contents of one database disk can be physically copied to another preformatted, similar disk. Disk routines are still another. The disk-handling routines provide both a housekeeping interface for addressing a particular disk spindle and also a videotex interface so that a chosen disk file can be accessed irrespective of the device on which it is loaded. The directory services assure directory access and updating for the database and user file services together with allocation of disk space. And the adopted protocols assure efficient data transfer between the storage media and the visual displays. Figure 10-8 outlines the record and field formats, and Fig. 10-9 presents the system indexes.

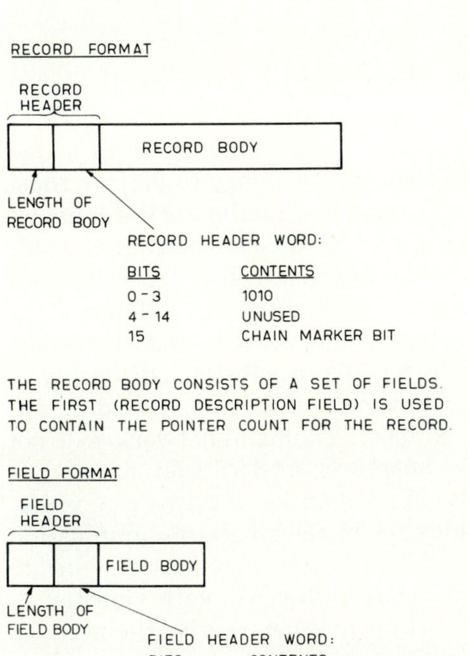

Figure 10-8 An outline of record and field formats for videotex pages.

```
INDEX SECTOR

EACH SECTOR CONTAINS
      1 WORD CHAIN ADDRESS
      1 WORD PARENT ADDRESS
      1 WORD NUMBER OF ENTRIES

EACH ENTRY CONSISTS OF
      - 1 BYTE LENGTH OF KEYS, AND
      - A NUMBER OF DISC ADDRESSES


THE INDEX IS TREE-STRUCTURED
```

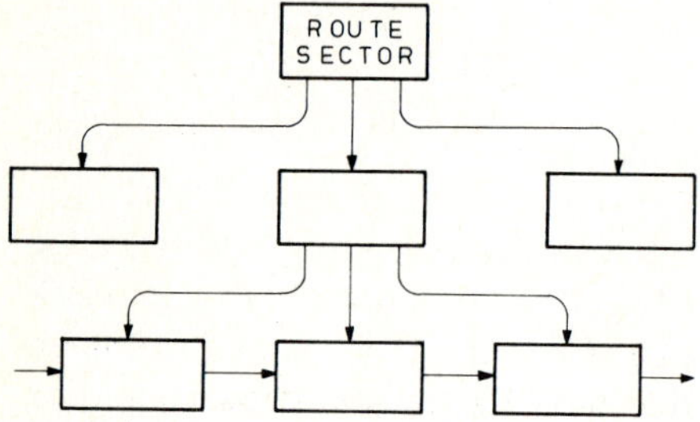

Figure 10-9 System indexes for menu selection and the index or routing tree structure.

Accounting for System Limitations

The service that any message delivery system provides is to accept messages conforming to a specified format, attempt to deliver those messages, and report to the user on the success or failure of the delivery. This service is supported in the context of a transport mechanism able to carry the messages from origin to destination. It involves relaying a message through several intermediate nodes utilizing different communication services or, alternatively, a single communication system.

In the former case, it is possible to have a substantial variety of messages in terms of format, data structure, and the like. Also, different communications disciplines may be used. Then the network will not only have to assure routing, flow control, error detection, correction, and so on, but will also need to observe a number of conversion procedures which add so much to complexity. In the single-communication case, those requirements are simplified, but there are specific rules to observe and, therefore, constraints. That is the case with videotex.

Furthermore, access to the VDB brings into perspective the need for understanding of the system's limitations:

1. *Interrelation.* ("Data with data" as in personnel records: age and experience, experience and living location, people and age). Interre-

lation is something we can easily do with realtime, not with video-tex.

2. *Format.* The format is 40 by 24, and so it is smaller than that of the classical RT video terminal—the more so if contrasted to with hard-copy output.

3. *Transmission speed.* Because of using voice-grade lines, videotex is limited to 1200 or 2400 bps. But that is true of RT also.

4. *Static information.* The information in the VDB may be updated regularly or only once per day. This introduces the concept of real enough time. If the VDB is updated, the frame looked at may be static (Prestel) or dynamic (the London Stock Exchange system). Typically, however, some 60 percent of the data is static on a daily or weekly basis.

Viewdatabase design and frame development should also account for the so-called orphans—pages without parents. With some systems, such as the Philips system, that is impossible. With others, such as Prestel, it is possible but illegal. With still others it is acceptable. The latter are more flexible, but they also call for more precise design solutions.

So in designing frames, we should look for fields which may be changing faster than others. If the software supports it, local editing of fields on a screenful will improve the overall upkeep, but we must follow standards and rules applicable to all users. A typical example is stock exchange information. From company identification price/earnings to ratios the data is fixed or slowly changing. But the prices are dynamic and steadily changing when the market reacts.

Some of the limitations are temporary. For instance, the issue of interrelation can be solved through personal computers working with videotex. Indeed, personal computers can pull out of a public system the right information and use it intelligently. The facility will, however, lead toward dividing frame design between data and text to be accessed as they stand in the VDB and those designed to enhance a personal-computer-supported database.

The basic decisions to be made about frame design should, therefore, focus on:

1. How the user gets to the information

2. Where the data is stored and manipulated

3. How much room for growth should be provided

4. What sort of flexibility we wish to build in (for example, different routes for different users)

5. How well we understand capabilities and limitations and make use of them

A basic design decision regarding the access mechanism is whether the overall approach will be top down or bottom up. The former tends to be more favored. It is better to take a top-down design approach, but it is also necessary to periodically check fitness and interact with bottom-level solutions (infopages).

Another basic design decision regards the human organization behind the videotex mechanism. This may involve:

1. Establishing the database administrator (DBA) function

2. Providing text editing procedures

3. Deciding on the use of structural approaches and standardization

The DBA must study and establish priorities, authorization, and authentication. It is wise to include the authentication keys in viewdatabase design at the beginning. It is also advisable to provide routines able to take charge of the registration of DB users as the application proceeds.

Typically, both the routing frames and the infopages will be built offline; but after establishing the offline overall structure, the designer should both project and check the links. The designer should also avoid changing or eliminating links online. The system will reject the change.

The main trouble spots we can encounter are:

- Removing an established link (cross-index)

- Failing to protect the information from unauthorized read/write

- Failing to reflect into hardcopy image the 40 × 24 softcopy. Two different presentation standards lead to confusion

We have to learn how to write in a small space and avoid overflow and also how to pick up and emphasize the most important points of long reports. (It seems difficult, but it is not.)

As to design, we should avoid overusing flash and color. We should watch the combination of colors and use some standard colors for standard data (page number, heading, and so on). Further, we should not overcrowd the page. Above a certain level, the more information there is on a page, the more difficult the page is to read.

A golden rule is to make the page action-oriented. Videotex pages are intended for management, and we should use color and/or flash for exception reporting. Just as important is the proverbial long, hard look at evaluating pages and databases. Evaluating and testing is a lifetime job. The designer should take time to examine the database that has been prepared and think how to use it better and how to improve it. What is the purpose of the Data? For top management? For middle management? For operating people down the line? Is reporting now done through printouts? How often? Daily? Monthly?

11

Thinking of the End User

"A memorandum is written not to inform the reader but to protect the writer."

DEAN ACHESON

The best design of an information system is achieved when we *first* look at the end user functions and *then* build up the system components into a working ensemble able to discharge those functions. This is roughly the opposite of what most system analysts have always done, yet it is a valid principle whether we talk of videotex or of more classical information systems.

What will the end user be doing when he starts operating on the system? His initial step is to switch on the TV, call the videotex computer, and on his keypad (Fig. 11-1) key in his identity number. Next, he keys in his passcode. If the identity number or passcode is incorrect, the system will allow two more attempts before disconnecting the line. If both are correct, the main routing page will be displayed on his screen.

To perform operations, the end user should either remember or have access to the standard functions (Table 11-1). He must be instructed on how to access the *user guides* supported by the machine for prompting and assistance purposes. (The latter is a straightforward operation. In 2 days people can be trained to become *editors*, and they can be trained much faster to be efficient users.)

The user's reading habits also must be reflected in design considerations. The average text page tends to be at the 600- to 700-byte level (as indicated by statistical evaluation). The exact size depends not only on the application but also on how much color, graphics, and other features are employed. Yet in database design calculation, it is advisable to use 1000 bytes per infopage. Some systems permit text and data

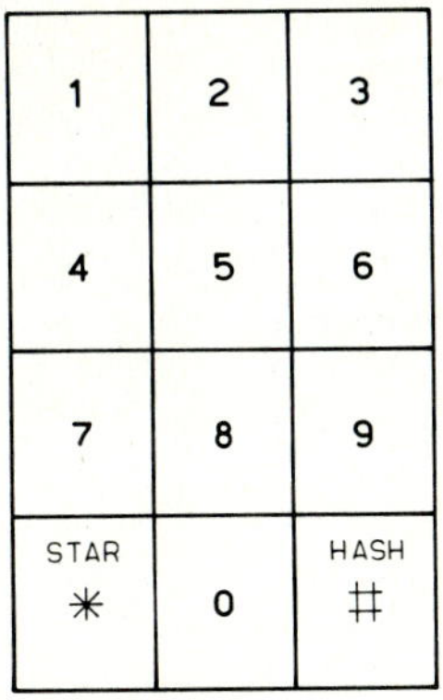

Figure 11-1 The simplest means of accessing the viewdatabase is the keypad, well known from touchtone phones.

Table 11-1   Standard Functions for Interactive Videotex
All functions start: *0...

Function	Argument
*0#	Display main index (main routing page)
*00#	Retransmit same page from VDB
*01#	Help (basic functions)
*02#	Help (access functions, such as *...*, *N#, N, N#, keyword#)
*03#	Reserved
*04#	Rotate frame being viewed
*08#	Switch user ID (UID)
*09#	Log off

For Page Selection	
Frame number	Go to the next frame of an infopage
	Go to any frame of the infopage viewed
*Page number	Go to any page or frame (part of multipage)
*	Go back to the previous routing page looked at
Keyword	Go to a page identified by a keyword
*Correct instruc.	Correct or change an instruction

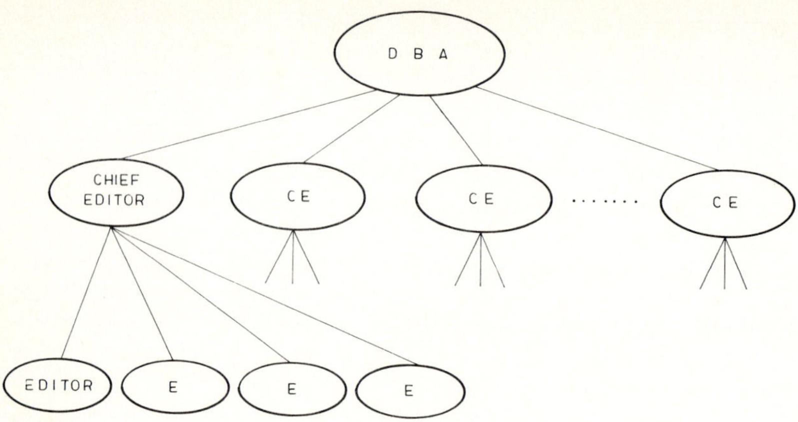

Figure 11-2 A hierarchical structure reflecting editing responsibilities for the viewdatabase and its contents. Reporting is to the database administrator.

compression, but that is not necessarily the best solution. It exchanges low-cost storage space for high-cost executive time.

As we will see in the appropriate chapter, infopage editing will be in the hands of the editors and any authorized person with an alphanumeric or editing keyboard may carry out the function. However, for coordination purposes it is wise to keep routing editing in the hands of the DBA alone and to restrict page editing to a limited number of people.

Particular attention should be paid to authorizing people to edit (practically, *write*). The overall control should rest in the hands of the database administrator, supported by *chief editors*. These latter also should be kept to a minimum, and so too the editors who report to them (Fig. 11-2).

Training the Specialist

Editors should be properly trained. For any new insert, they should be instructed to identify the type and identification of page, insertion into the VDB system, retrieval, update, and end use. At no time should the editing operation upset the system's menu capability. For instance, the latter requires particular attention to index page insertion. An editor's response frame will display the image as shown in Fig. 11-3.

Furthermore, though the writing of the infopages necessarily reflects the application (and we will return to this subject), the design of the routing pages should be consistent, homogeneous, and transferable from one application to the other. The same thing is true of the multipages. The sequence starts with the routing pages. Once an infopage is arrived at, it automatically becomes a *multipage*. The displayed text identifies the frames attached to it with the exception of the first three lines, which are for housekeeping.

Though identification is supported on the title page in a table-of-contents manner, addressing is supported by the system software. Within the multipage we can use an XX# call and then return to the XX page. This is the browsing capability. The system also remembers the last *four* pages addressed and, through *#, gives the preceding pages four times downline. Furthermore, multiple-to-one relations are actuated through crosslinks. This is important in:

- Providing a consistent image to everybody

- Updating only once

- Using reserved pages to support this purpose

Reserved pages usually reside in the system part of the database. This does not apply to a multipage. Furthermore, reserved pages cannot be deleted by the user; they are delete-protected.

Let us review this discussion of the multipage. Many users build up a table of contents. It may be alphabetical or organized by another criterion (subject, destination, and so on). However, it is necessary to establish cross-indexing able to update dynamically all index pages by updating *only one* of the lists and then using cross-indexing. As with all data entry operations, the big secret is to have the elements in the database *only once* and use them many times. That is the best way to

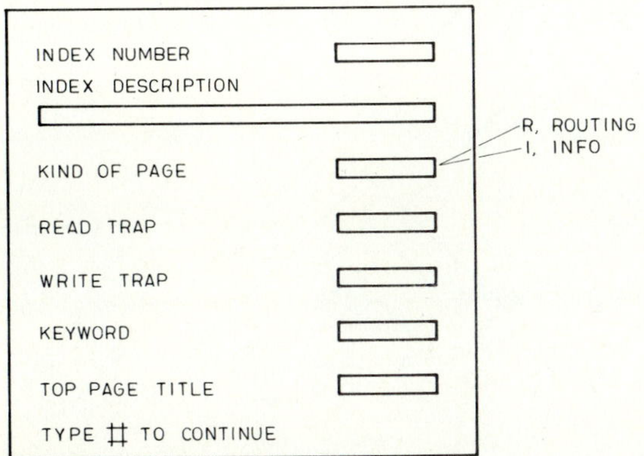

Figure 11-3 Editors should have available formatted response frames to help them do a better job. An example is given here.

assure efficiency, consistency, and reliability in the database and to do more for less.

The designer will be well advised to avoid multiple copies of information in the database, try to keep the database to the minimum, and, rather than work harder and harder, work more wisely. The principles are simple, and they are quite similar to those of network database design.

The Trap Function

We referred to trap in Chap. 9. No system can work without a properly designed and actuated protection mechanism. If it is lacking, the equipment and its software will be useless for management purposes. Protection should work two ways:

- In obtaining system entry rights (we have mentioned ID and PIN personal identification numbers)

- In assuring the access to the pages of the sub-VDB for closed user groups and other purposes

The latter protection function is carried out with the *trap* code. Let us look at an example. Say that page 7356 is reserved to user profile with code 100. The page will be accessed by user X. If his user profile does not

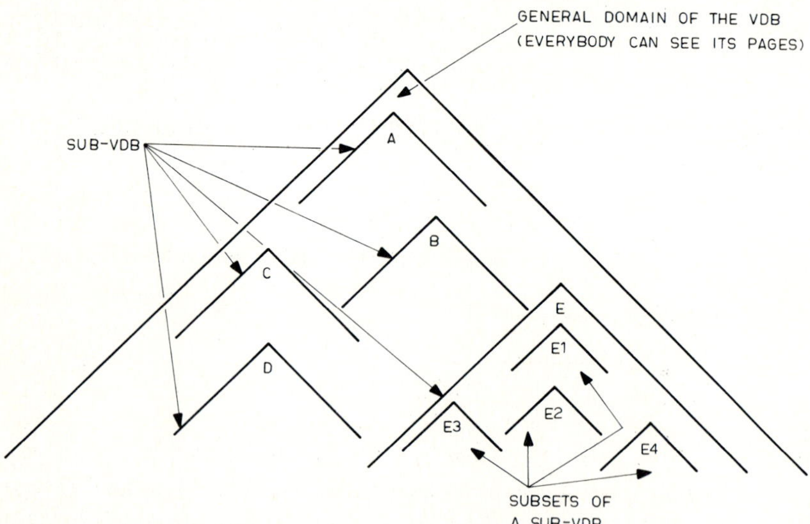

Figure 11-4 Because of the closed user group facility and the resulting security and protection, the viewdatabase can be subdivided into subdatabases, each of them managed by its owner.

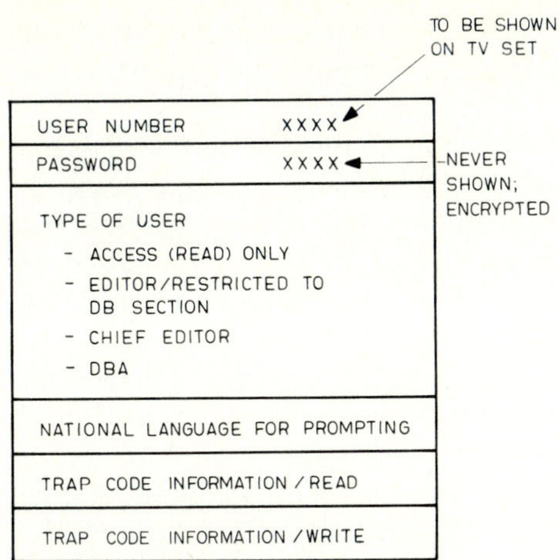

Figure 11-5 This typical protection sequence involves a user number and one or more passwords. Some banking systems assign to the end user the ability to identify his own password(s). The password is for entering the system, not for protecting a given page. Pages are protected by trap codes.

have 100, the system says there is no page like that, although the page may exist in the VDB. This brings into perspective a whole sequence of operations, with the definition of the subdatabases being the most important of them all (Fig. 11-4). Typically, these sub-VDBs will be protected through the trap code. Also, each sub-VDB can be further subdivided into sections each of which is further protected.

This approach assures a valid, chainlike protection sequence as demonstrated in Fig. 11-5. The sequence will basically involve a user number identification and a password. The two are stored on disk and are needed to enter the system. To protect the page we must use central memory access to reach:

- Trap/read

- Trap/write

For instance, one of the available alternatives, the Philips videotex system, supports up to 240 trap codes. They are available to the DBA for allocation according to needs. Another 15 trap numbers are reserved for the system database. Though the same code can support read/write, usually we should assign different trap (protect) numbers to the *read* and the *write* functions to make the system more flexible. However, different pages can be supported through the same trap number.

Access standards are under development by international bodies also. The new CCITT recommendation for a videotex standard introduces a control character called inquiry; it allows the terminal to supply additional information prior to granting access. *Inquiry* and *PIN* thus help differentiate between station (terminal) and sender/receiver. A further distinction might eventually be necessary to help identify the *author*.

The following can thus be said in regard to the safeguards to be used: The design of a VDB must fully account for the areas of protection. A closed user group has to be defined *prior* to VDB design, and we should always reserve spare trap numbers for later use. If the trapping of lines is a particular worry, it is possible to insert a scrambler/descrambler unit between the modem and the line. Should that be done, we should be very careful to use consistency throughout the system.

The existence of crosslinks in the page system poses a particular trap-setting challenge. Crosslinks between different user trees must have common traps to avoid a breach of security. Furthermore, to establish use perspectives properly, system analysis should consider as distinct entities the user groups with read-only capability and the edit groups. Both must be examined as to responsibility and supervision. Information which is valid to more than one group of users should be placed under common management. Coordination should be provided through the latter, and proper, consistent design requirements should be met in a similar way.

Facilities Available to the Editor

We have said that the editor is able to *write* into the VDB; if a person has write privileges, he is automatically an editor. However, we can further distinguish between editors by classifying them as normal editors and editors with trap-setting privileges. The limits are imposed by the DBA of the system; for he controls both the structure of the VDB and the editing of the data.

The normal editor will not know about the specific traps, because they are run automatically by the system (after being established by authorized personnel). There is also a user's response frame capability for electronic mail (videotex) purposes.

The electronic mail function is in parts:

1. *The editor can define a response frame for an end user.* In this formatted approach, the editor preestablishes fields and inserts the page in the DB.

2. As an alternative, there is the possibility that *the editor incorporates blank pages to be employed by the end user (equipped with an alphanumeric keyboard) at the latter's choice.* The alphanumeric keyboard offers degrees of freedom but also is more demanding of the software. The system checks if the information is correct syntactically (not sequentially). Quite evidently, the vehicle of response frames can be used for message switching and electronic mail.

3. *The editor can also create an applications subsystem by ingeniously manipulating the functions supported by the system.* Consider, as an example, a stockmarket application. Say that, in VDB design, the multipage is composed of five frames. We pick up one frame and call it:

$$* \ \underbrace{04 \ 1}_{\text{Rotate}} - \underbrace{99}_{\text{Timer}} \ \#$$

The system will start at page 1, stop XX seconds (as stated in timer) and then go to the next page. By using the rotate frames function, we can get the updated frame on the video. The same function can be used for public exhibition and other demonstration material. It is, furthermore, possible to cross-index the same pages from two trees (Fig. 11-6) and thus update a single stockmarket multipage.

The assignment of the facilities described above tends to divide the

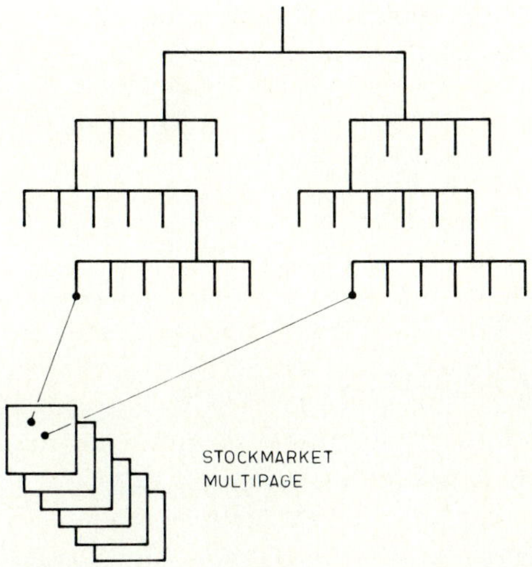

Figure 11-6 Cross-indexing a stock market multipage. The cross-indexing feature is very helpful to the end user and also to the person responsible for designing and upkeeping the viewdatabase.

day-to-day user population into two large sets: those with *access only* and those with *editor privileges*. As Fig. 11-7 outlines, the former have read-only capability; the latter can be divided between the *page* and *structural* editing capabilities. Still further on the VDB management side are the trap editors and the DBA.

To recapitulate, the use-authorization levels are:

1. Access user (read only)

2. Editor (limited by trap code)

3. Editor with trap-setting privileges

4. Database administrator

Supporting the access functions is a range of software and procedural routines. Consistency controls are most important. If two editors try to work on the same page simultaneously, the one taking control has access to the page. The other is informed (by the system) that the page is worked on at the moment. Therefore, he cannot alter it. For journaling purposes, the system addresses a given page and assigns the first editor the editing rights on that page. That is why any other editor addressing the page can read it but cannot write on it. Journaling also supports recovery/restart.

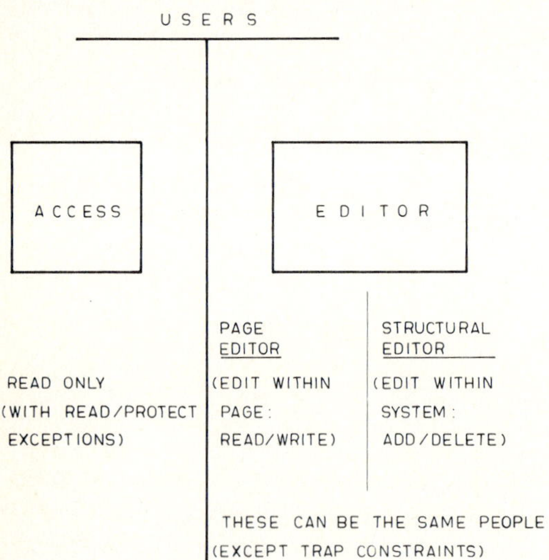

Figure 11-7 Depending on the purpose for which a videotex system is implemented, it may be advisable to divide clearly the access responsibilities by editors and end users.

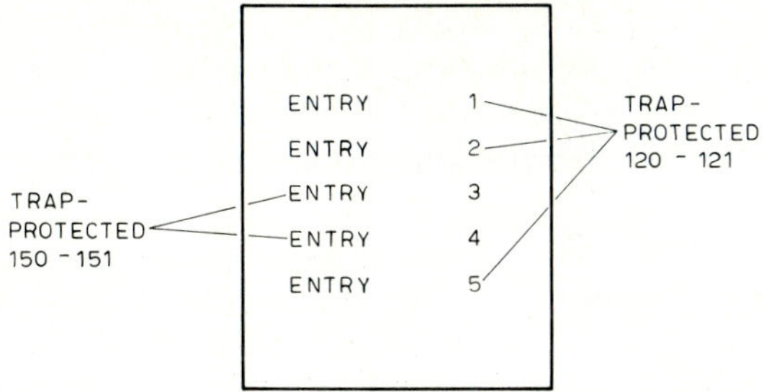

Figure 11-8 A routing page. It is advisable to protect such a page by means of a trap or other mechanism. Unauthorized changes in the page can upset the whole system.

An authorized editor can edit (change) the routing page, but routing pages can be trap-protected (Fig. 11-8). It is wise to use the trap feature to assure that one organization does not enter the editing structure of the other. Some videotex systems time the length of a session. Others have no clock to limit the time a page can be manipulated by one editor and so become unavailable to another.

The accounting system to which reference has been made principally applies to the end user. It employs units to which monetary values can be attached and so be made universally applicable. Switches from user identification (UID) are possible through a software facility of log-off from old UID, and log-on with new UID, without disconnecting the system. The code is *08#. Basically, however, the assignment of user numbers should follow a taxonomical approach, so that, by looking at the user number, it is possible to say who the user is.

The structure of the user identity system, the routing pages, and the VDB design at large, should be the subject of meticulous planning. As with every online information system, changes in database structure mean conversions, and they result in delays, costs, and the spoilage of resources.

Graphic Presentation

Like Cato in the Roman Senate, who at the end of his every speech hammered into the mind of his colleagues "Carthage must be destroyed," I will never tire of repeating that *management presentation must be predominantly in a graphical form.* I said that in the discussion of decision support systems. Let us now see how we can implement it.

First of all, the objective is clearly stated in Fig. 11-9, which is an extract from a slide projected in a lecture at the Harvard Business

School. A forward-looking company substituted, for its 300-page report to the board of directors with its long (and traditionally unwinding) tables, 18 graphs and two short tables that gave senior management what was needed when it was needed: order of magnitude and direction of the crucial indicators.

At Gould Industries, O'Hare Airport, Chicago, I had the opportunity to observe an interactive application of the same principle: Through specially designed terminals the members of the executive committee could call (and present on a wall-sized panel) graphs and data through direct access to the VDB. That is the basis top management should have for action.

Whether through videotex or more classical systems, computer graphics will play a greater role in the future as flexible software is in place. Graphical applications, however, presuppose that an in-house database provides information for studies and subsequent implementation. Both internal and external information can be vital. The latter, for instance, can have to do with banking habits, per capita savings accounts, population density in a prospective area, and median levels of income.

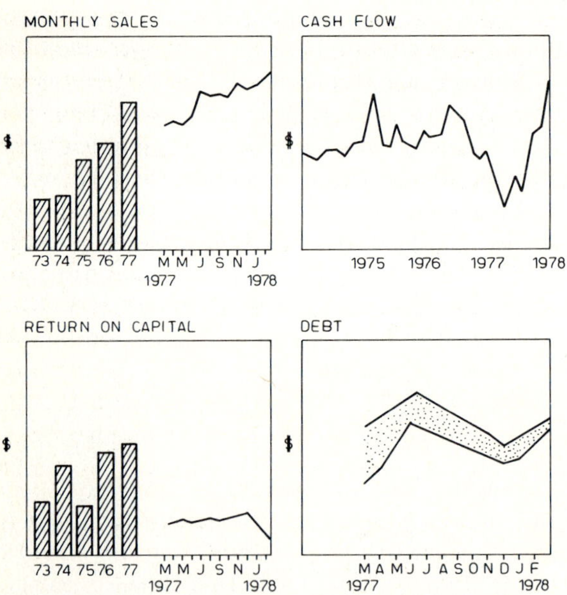

Figure 11-9 For its 300-page report to its board of directors, a forward-looking company substituted graphs and order-of-magnitude tables. These graphs are selections from the presentation.

One of the pillars of success is the fast turnaround that the in-house graphics capabilities should support. Videotex presentations must be prepared for top management use often within hours after the data is analyzed, and the equipment can provide the graphics needed to meet such a deadline. Color should be included for exception reporting. This is the wise course to take, and current attention is focusing on it.

To gain experience, a company may start with graphs through an editing terminal, but in a second phase graphics should be converted to an online application. Two possibilities of a suitable package come to mind. One is a software solution that makes available, without the need for programming expertise, a graphical capability which permits customizing and tailoring through parametric specifications. Though it may be memory-consuming, its use in night shift for the preparation of videotex management graphs minimizes that limitation.

The second possibility is an easy-to-use graphic terminal. The machine can be quite simple: An Apple computer which, together with a four-color graphic tablet, permits the preparation of four-color graphs in about 2 min, stores them on a floppy disk, and can be connected to mainframe for graph integration.

There are also home-made possibilities. Data processing may choose to write a conversion routine through its own specialists. For instance, a timesharing routine whose output is character-oriented (not alphageometric) could be used; it is available on many mainframes. Even if the character output is not necessarily acceptable to the VDB because it is expressed in a different code, there are two ways to handle the discrepancy:

1. Convert the character code output to alphamosaic

2. Do a routine to convert the tables, prior to output, into videotex code

The timesharing printed graphics can serve our purpose if we follow either of those paths. Given that, say, an alphamosaic videotex system supports a 24 by 40 matrix (say 20 by 40, given ID and follow-up) and that semigraphics are drawn on a two-column, three-row basis per character, we must convert to a matrix level (Fig. 11-10) of:

$$
\begin{array}{ccc}
20 \text{ X} & & 40 \\
\times 3 \text{ X} & & \times 2 \\
= 60 \text{ X} & = & 80
\end{array}
$$

Hence, our matrix will be 60 x 80 and we can comfortably support that level with the understanding that it makes feasible all typical management graphs—histograms, bar charts, trend lines, and ranges—or combinations of them. The limitation of the method is its single color—because all cells must be the same color—but we can edit in color at the editing terminal if necessary.

The computer people could use a data tablet (or digitizer), an excellent device for translating existing artwork, such as drawings, into the digital form required by the computer. Freehand drawing can be input more naturally with a light pen, and the cursor control keys are an effective input when horizontal and vertical alignments must be followed. Anyone who has worked with computer-supported graphics will appreciate the ease with which a graphics database can be established, edited, or updated.

Furthermore, different technologies are now merging, and videotex is also successfully used for animated graphics by looping alphamosaic pictures to create a continuous stream of images. The character of the motion determines the technique used to express it numerically. All smooth, precise movements can be expressed as simple algorithms. Loops can produce richer, choreographic movements betwen the chosen frames. The technique must be used with care. Experience is necessary to know what is making a move better or worse. With alphageometric approaches, light pens and data tablets can sample points in two dimen-

CHARACTER LEVEL ALPHAMOSAIC DESIGN

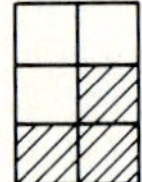 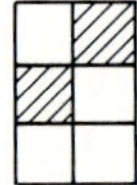 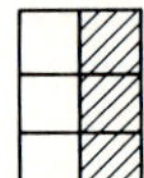

CODING A MANAGEMENT GRAPH FOR VIDEOTEX

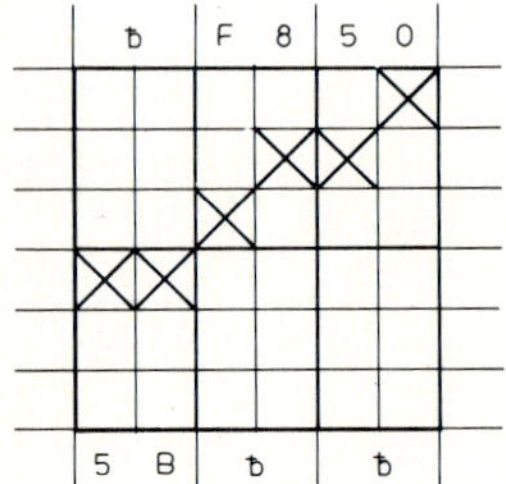

Figure 11-10 An approach to converting to alphamosaic figures for the production of histograms, bar charts, and so on.

sions at a very rapid rate and permit subtle movements to be digitized directly.

These approaches can be used both by amateurs in the graphic arts and by professionals. Professionally, since the computer is not fast enough to draw the images at 20 frames/s, a color motion test is filmed directly from the CRT, frame by frame, by using a 16-mm (millimeter) camera under computer control. It is routine for the computer to expose 4000 frames unattended.

Control must, however, be exercised because this graphic art is still in its infancy. When the motion test is projected, all aspects of the computer graph must be examined carefully. Management graphs require much less skill than the high-resolution plots corresponding to the frames of, say, a motion-picture-animated graphics presentation. Yet, also in the management case, better care is needed if not only softcopy but also hardcopy must be produced.

Technology provides the tools; but to benefit from them, we should assure the infrastructure. The lack of procedural integration will handicap the effort to modernize the management system. Delays in the transmission of information for decision purposes will kill interest in its use. Data that is too compartmentalized will reduce the benefits and that underlines the need for integration of the data assets at our disposition. Once integration is effectively carried out, the following areas will be open to the use of graphics within the banking environment:

- Performance records
- Plan versus actual
- Forecasts
- Trends
- Strategic planning charts
- Cost of money
- Economic accounts
- General accounting
- Management control information
- Financial information
- Market penetration and market share
- Characteristics of the client base
- Management of the client base

- Investments (by type, interest rate, and maturity)
- Evolution of deposits
- Monthly developments by client class, branch office (BO), and type of deposit
- Risk analysis
- Loans, exploded by class, interest charged, and industry
- Average, max/min time, and deviations in handling loan applications
- Variation in decisional results (and delays) by branch office
- Profit margins by type of loan, spread, and industry
- Commercial paper statistics
- Delays in the payment of obligations
- Exceptions to good practice
- Total business, broken down by geography and type of client
- Internal organization
- Organization memos
- Internal norms
- Personnel records
- Legal communications
- Operating documentation for computer operations
- Diagnostics and reliability statistics
- User statistics

Though several kinds of information are common to top management and branch office management, reporting levels and mode of presentation can be quite different as to detail and critical evaluations and comparisons. These are useful guidelines to keep in mind because they help avoid redundant effort and erroneous approaches which must eventually be corrected at the cost of time and money. Also, they are as applicable in banking as in any other industrial activity.

As far as management graphics are concerned, the rules are fairly common. Attention should be given to making both the graphical and the tabular presentations in ways that are appropriate to the levels of management to which they are addressed. Tables 11-2 and 11-3 exemplify that observation. The first compares the programmatic interfaces

Table 11-2 Programmatic Interfaces of Videotex-Supported Reports

	Top management	Auditors	Functional management
Economic account:			
Liquidity	X		X
Investments	X	X	X
Property	X	X	X
Portfolio management:			
Financial	X		X
Inventory		X	
Foreign operations:			
Spot	X		X
Time commitments	X	X	X
General accounting:			
Day-to-day		X	X
Management acc.	X		X

Table 11-3 Possible Use of Videotex-Supported Reports

Type of information	Top management	Auditors	Branch office management
Client base	X		X
Current account:			
Reports with clientele	X	X	X
Exceptions on good practices		X	
Profits by client position	X		X
Costs	X	X	X
Branch office Profitability costs	X	X	X

of four major areas to the needs of top management, the auditors, and functional management. The second outlines the possible usage of some reports, including their application to BO management. Figure 11-11 relates (in a banking environment) the functional databases to the management-oriented VDB.

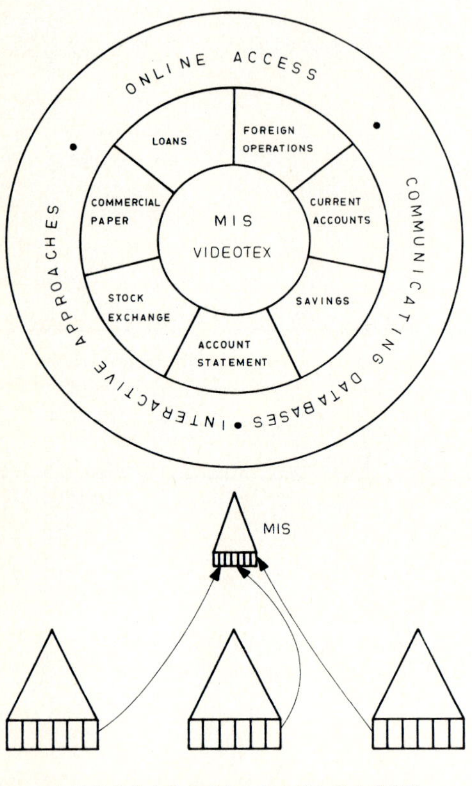

Figure 11-11 The approach followed by a financial institution in relating mainframe and mini-based databases to an MIS viewdatabase.

12

Converting to the Viewdatabase

*Never insult an alligator until you have
crossed the river.*   HULL'S WARNING

The chapters which dealt with VDB organization should have left no
doubt in the reader's mind that a significant amount of work must be
invested in order to produce a rational and solid VDB. Yet no matter
how much this work may be, it is certainly only a fraction of what would
have been necessary to reorganize, revamp, and reinstate the total
database requirements within a fairly complex operating environment
such as that represented by the typical industrial organization with
years of experience in data processing.

Let us return to the three layers we described in connection with
decision support systems:

- Top management

- Middle management

- Transactions

We said that, two-by-two, these layers have things in common: transactions and middle management share the detailed database in the way
it was built throughout the 1970s. And as we all know, that database
was bulky and crammed with detail. When it comes to decision making,
the trees can hide the forest.

Perceptive organizations have seen this point. The Bank of America
has a rule that 80 percent of the data born in the periphery can die in
the periphery; it does not interest the center. Gould Industries also
applied that principle. The different divisions of the company keep their

DP installations as they stand, but every night they extract management data (mainly summaries) and send them to headquarters through public lines.

The reference to decision support services, text and data handling, and graphical presentation rests on this simple mechanism: The data updates the VDB and makes it actual and accurate for management purposes, and also easily accessible. Citibank's IAMIS (Internal Accounting Management Information System) is another example along the same line, and at General Motors it is said that Ed Cole, its former president, personally collaborated in devising the most effective means for management presentation.

In this sense, the organization can develop a distributed information system that takes computer power where it belongs—to the workplace—while management keeps control through its dedicated MIS minicomputer (Fig. 12-1). The challenge is in creating the proper homogeneity and in extracting from the day-to-day database the information elements necessary for decision support purposes at the top. This work must be automated. It cannot be handled for long by hand.

Organizing the Viewdatabase

Creating a new database concept, such as that implied by the VDB, is one of the opportunities which do not come too often in information systems life. As Chap. 3 demonstrated, we must get properly organized,

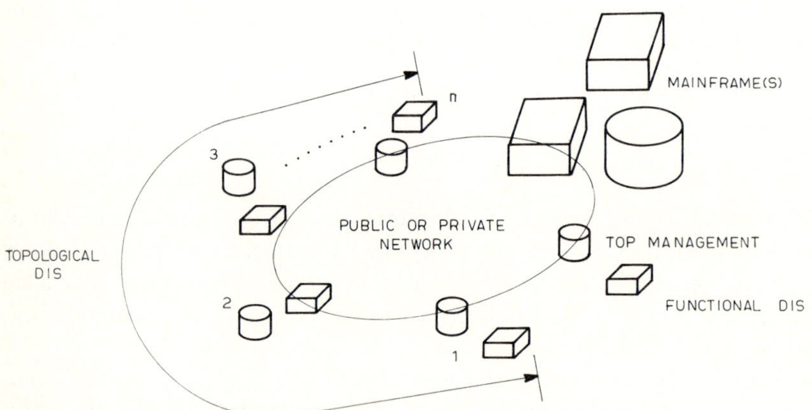

Figure 12-1 The top-management-dedicated computer, and other management systems, should not be confused with the aging central mainframes.

and that means looking both at the DBA functions and at the end user VDB requirement in a rational and documented manner.

The DBA reference (and the management of the VDB at large) suggests the creation of a *database* that properly identifies the infopages which constitute the VDB (encapsulated information) and their possible use well on the way to cross-indexing. This calls for well-established steps:

- Name of the total VDB
- The assignment of passwords to gain access to the system
- Bootstrapping capabilities
- Name and address of the DBA in charge of the VDB organization
- Title of main index
- Establishment of a page header policy (left-side top line of the info-page)

To be successful, we must put to work the proper human resources both quantitatively and qualitatively. It takes some 3 to 4 months of concentrated work to design a good VDB at the 100,000 frames capability level. Once that is done the right way, applications perspectives open up. Given a solid, logical organization, the change of physical support, for instance, will pose no problem. Manufacturers assure VDB portability from, say, 10- to a 20-MB disk.

Database knowledge is, then, a prime requirement, and serious work will typically involve the following procedures:

1. Establish the system perspective.
2. Provide for database integration.
3. Look after database partitioning.
4. Elaborate text and data structures.
5. Assure consistency.
6. Guarantee updating capability.
7. Look after integrity of use.
8. Guarantee the one-storage, many-users principle.
9. Make sure that security is strictly observed.
10. Provide for database maintenance.

The analysts to be entrusted with database design should read the documentation carefully, proceed with the analysis the right way, and make sure that all facilities supported by the computer are properly used, and not just 10 percent of them. As in all sound system analysis procedures, they should start designing on paper, refine their design and simulate prior to installing the database software, preoperate the "empty" database, and train the operating people in database use. They should establish a list of unit keywords, put into the VDB some textual information to get a feeling of how the VDB works, and start experimenting with what they have up to that point.

A particular challenge is the organization of the routing functions. The job starts with the construction of the routing tree. While viewing the routing pages under development, the analysts should simulate the operation by selecting an entry on the page. The simulation must take into full account the action of the end user while viewing an information page. The user may

- Select the next frame
- Select any frame
- Rotate the frames

To properly structure the access and editing capabilities supported by videotex, the analysts should develop a flexible mechanism able to create new routing and information pages, copy existing information pages, generate crosslinks to existing pages, change entries on routing pages, save infopages on external files, and delete pages. Those requirements help to clearly identify the procedural steps it is advisable to follow in the design phase.

1. Project the VDB structure.
2. Create the menu selection tree.
3. Assign direct and keyword access capabilities.
4. Provide authorization and authentication (passwords and trap) facilities.
5. Assure subsequent database maintenance.
6. Keep in mind the database administration function.

These are fundamental steps to be undertaken prior to the more detailed responsibilities such as deciding on the use of the accounting routines. Typically, for billing and statistics purposes accounting will be done at two levels: elapsed time and total number of frames looked at. Another fundamental step is *save*. It can be taken on the videotex equipment through either two disk drives or tape drives.

The Logical Structure of the Viewdatabase

The menu selection function supported by videotex is instrumental in making the system user-friendly. In turn, this function sees to it that VDB design is hierarchical. Figure 12-2 demonstrates the logical organization of the database. Two sections must be distinguished:

- The system VDB
- The user VDB

As we will see, the latter must be divided into a number of sub-VDBs each protected through trap numbers to guarantee a closed user group (CUG) operation.

The first digit in the system VDB tree is 0. This leaves the range 1 to 9 for the first digit of the user VDB. Each of the main VDB sections will involve the three key elements of which any VDB consists:

1. Index pages
2. Infopages
3. Functions

The functions are transparent to the end user. They guarantee the storage and access capabilities and a number of other features such as security and protection. They also make feasible the management of

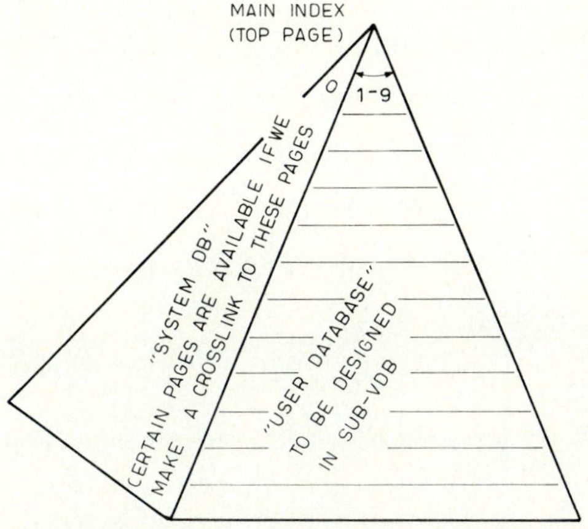

Figure 12-2 Logical organization of a viewdatabase distinguishing the system-oriented and the user-oriented VDB. Always start at the top with an index page.

the basic paging structure: the main index page (at the top of the pyramid), the routing pages in the tree, and the information title frame, which opens up the multipage capability.

For security, protection, and confidentiality purposes, each *item* (page) in the VDB is protected by two keys: read and write. The keys support the entries (lines of the infopage). If entries are protected, they are not shown in the access sequence. The CUG facility is exercised at the routing level.

The logical organization of the VDB necessarily reflects the number of hierarchical levels which are supported. Here the system designer must be more thrifty than the manufacturer of the videotex software: Even if the system allows, say, 15 hierarchical levels in routing, a good rule is that design should not go (organizationwise) beyond 10 levels. When it does, it is a good indication the designer did it wrong and must start all over again. The more he works on the project, the more consistent he will be in indexing and cross-indexing and the more thrifty in the allocation of the different hierarchical levels supported by the VDB. Figure 12-3 gives an example of a routing sequence inside the user database. A typical design of a routing frame is shown in Fig. 12-4.

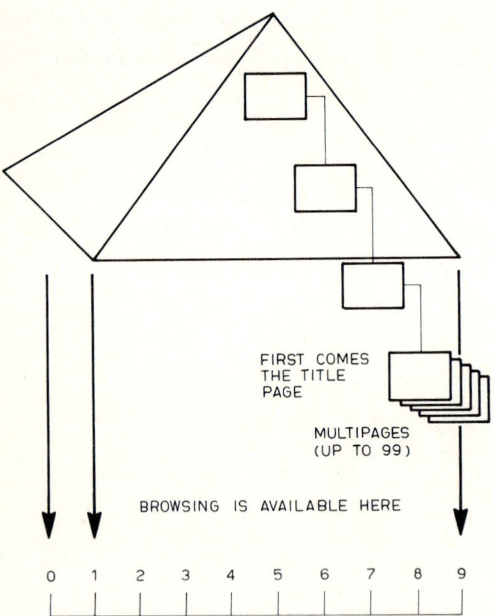

Figure 12-3 Following a routing sequence inside the user database.

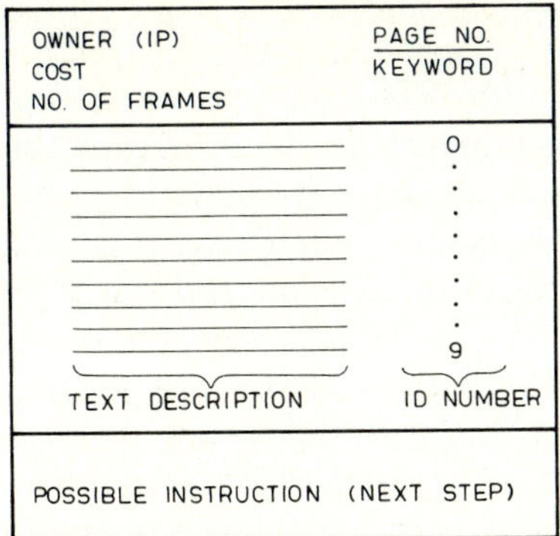

Figure 12-4 Typical design of a menu selection frame to be used for routing purposes.

Since studies have indicated that most people when looking at a page concentrate their attention on the upper-right-hand corner, it is advisable, when designing, to put in that corner the page number and keyword (alphanumeric). It is also advisable to utilize at best the facilities supported by the videotex software. For instance, to avoid losing time, a given feature permits automatic interruption of the display process (by pressing a button) once the sought-after line has been located.

Some software features can equally well be supplied by hardware to ease end user requirements. An example is the modem which automatically supports user number and automatic calling capability. Some business terminals also incorporate user number, and that facility permits the user to key in for access purposes only his PIN number, rather than both his sender/receiver identification (identity number) and PIN. Since a simple numeric keypad is a very easy access method for the management user (rather than a full alphabetic keyboard), it is advantageous to use a simple key combination such as: *0XXX to emulate alphabetic keywords in accessing a given frame in the VDB. These are refinements to be supported by the VDB software.

Other user-friendly solutions are the graphic tablet, mouse, joystick, and touch-sensitive screen. The mouse is employed in the most recently announced PC, and graphic tablets and touch-sensitive screens can be added to many microcomputers. A still more friendly access method is voice input.

Preparing the Conversion Routines

Having examined the logical structure of the VDB, the specialists applying themselves to the videotex project are ready to look into the subject of *conversion routines*. As we stated in the introduction to this chapter, such programs are necessary to assure that extracting and formatting the management infopages from the large database will be done automatically as required.

We will follow step by step exactly what should be done. It is, however, proper to state in the beginning that, in this phase of the operation, a number of services are to be supported and decisions made. One of the most vital is this: Will the videotex system we are designing be passive (simply display information) or active? An active system prompts the user to respond to information displayed.

Security and protection are the most vital of the supporting services. Multiple security levels must be assured. Pages of information may be held within a closed user group with which different access levels are associated. Before accessing the VDB, each user must be registered: The identity of his TV set is locked into the system, and he must enter a valid password. From then on, he has access to nonrestricted pages.

A long hard look must be taken when a user requests access to a restricted page. The system must verify that he has the necessary security clearance. Otherwise, access is denied and a warning message is flashed on the screen. Other design characteristics stress the system load: How much information does the organization want to hold on the system? How frequently will that information change in whole or in part? To how many people do we wish to make the information available?

Then we should examine with great care the logical and physical interface of the videotex with the mainframe. How will the information be captured for the VDB? Will the system be stand-alone, or will it be required to work in conjunction with a computer that the company already has? To what extent should we exchange information between the computer and the videotex system?

The systems work that is done must be first class: very thorough, well documented, and even pioneering, since we are at the beginning of the private user generation. It is advisable that the feasibility study on the conversion routines be organized into two volumes:

1. The general systems organization (macroanalysis) of the videotex interfaces to the mainframe.

2. The consolidated information system and user specifications

The first of these documents should identify the programs which constitute the subject interface. They can be organized into three main phases:

- Extract
- Other input
- Consolidation

The first phase is directly related to the files on the mainframe; and if the organization has adopted a distributed data processing environment, the analysis should be applicable within the DIS organization. Otherwise, it reflects reference to central files basically maintained through batch.

In the discussion which follows (and which is based on practical experience) the input is properly identified. "Other input" means mainly manual input, through editing terminals, as management asks the specialists to enrich the extract from the database with information not necessarily handled in its present form through computers. It also means graphic information to be extracted in large part from computers.

Graphics should be integrated after the graphical pages are formed. It is too complicated to feed graphics together with text and data pages while formatting; it is better to use a merge program. For the latter to work, it is necessary to provide coded information. The code might direct the carrying along of some structure which says that "this graph" is of interest to a specific text and data page. As an alternative, it is possible to reserve part of the structure for graphical information and provide cross-indexing to ease search by the end user or provide an instruction at the bottom of the text and data page: "If you wish graphs, key XX." In any case, it is advisable to make the search mechanism fully transparent to the user.

Here are some helpful rules to follow in *tabular layout:* Use color change horizontally, not vertically. The brain understands columns much better than it understands rows. The "color flip-flop principle, by alternate lines" saves blank lines, and space is valuable on videotex. One-color long columns are not difficult to read; alternate colors by column are most difficult. And try not to use more than three colors on one frame.

The existence of both a computer-supported extract phase and a manual input has led to the incorporation of a consolidation phase. Such a phase is a very good idea that new users should adopt; it helps specialize the software to one specific, well-defined task. The flow charts for the systems specification of the extract operation are shown in Figs. 12-5 through 12-16. In the application for which these charts were made, in terms of structured component parts, system design supports four key references through the different phases just outlined. Analysis is in four parts:

1. Initial input

2. Corrections

3. Output formats

4. Security

All the programs have been constructed parametrically to produce videoforms: daily, weekly, monthly, and on request. For instance:

- During a daily run, the typical extract program gives a breakdown of the stock figures by product and depot.

- During an end-of-week run, this same program produces two main reports. One gives a breakdown of the stock and sales figures; the other summarizes the stock and sales figures and shows totals per product.

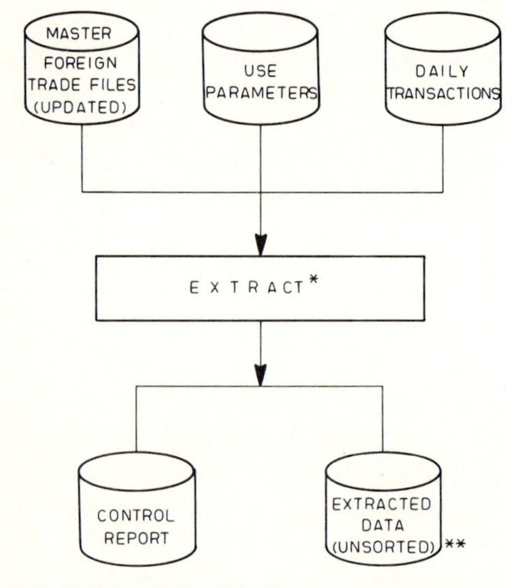

Figure 12-5 Data selection and extraction procedure.

Stock and sales data is read from the stock and sales consolidation file, as classically handled by data processing. The interface takes information from this file and formats it into screen images suitable for display by videotex. The image file under development is sorted into the required sequence by the mainframe, which adds the final page numbers and the color codes. Some fine programmatic interfaces also are involved. Examples are the replacement of the numeric 0 with alpha O so the screen is more readable and the display of negative figures in red on the screens for total and fixed stock columns.

System design has properly provided for the specialization of functions in order to keep the interface programs small and simple. One program, for instance:

- Assigns user and priority numbers for each screen being produced. This defines who can have access to the information on the VDB.

- Allocates an absolute page number to reference each page. This is

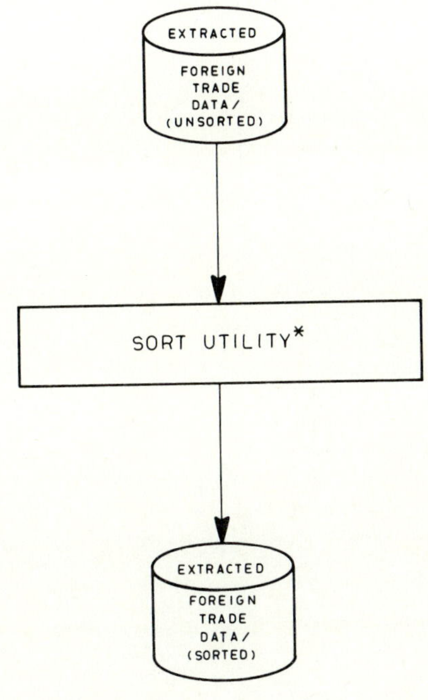

Figure 12-6 Sorting in preestablished order; creation of multipages.

necessary because the preceding interface programs have been assigned relative page numbers to be used in the editing procedure until the consolidation phase comes along.

■ Handles the color presentation.

Through exception reporting, the program supports a table with color control characters. These are maintained for each screen. They are extracted and positioned on the page immediately before the fields which are required in color for management reporting purposes.

Toward a Decision Support Environment

The mechanics are simple: The interface program makes a logical test and then sets the color code. But before color codes are assigned, a special action is required to ascertain whether the item under consideration is below a minimum number of weeks in stock.

■ The number of weeks cover in the column is examined and compared with the standard.

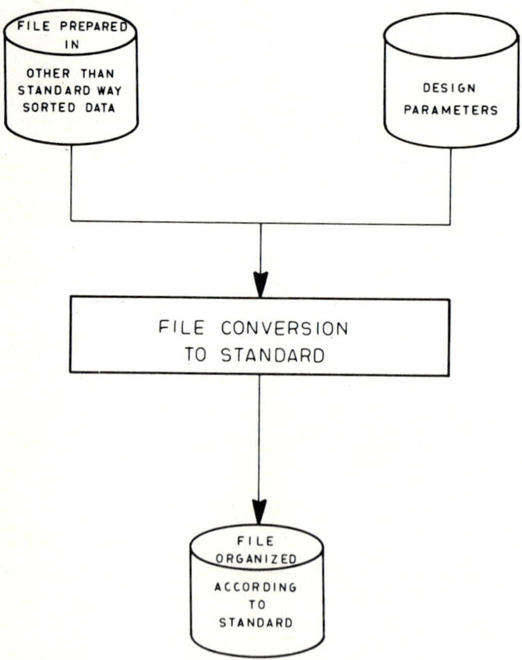

Figure 12-7 A parameter-driven program able to convert a file structure, say, of foreign trade data, to one homogeneous base.

- For other products, the same thing is done for the number of days cover column (to check if it is below a minimum number of days stock).

- For all products, the stock figures on each page are examined for negative values.

These are the prerequisites for exception reporting and its implementation through color presentation. The practice is very good: We can use a similar approach to color both the tables and the graphs which we need to develop for management reporting. The graphs might be created through routines on hand, *as if* to be presented by a character printer, and then converted to a videotex graphic code and colored.

The preceding are examples of the tailor-making requirements to which reference has been made. Though I am a great believer in the use of packages, the latter better fit an established procedural environ-

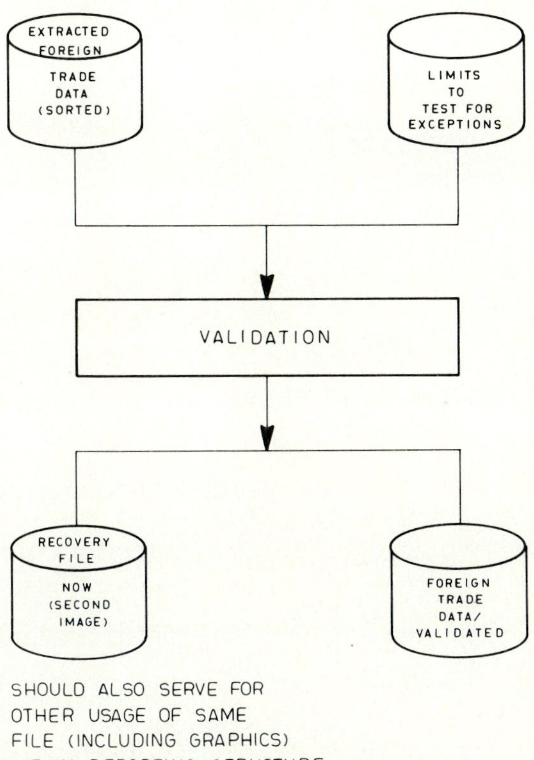

Figure 12-8 Validate data; assure conformity to ranges; insert color code for exceptions.

ment. With management information systems we are pioneering, and that necessarily involves a certain degree of personalization at the level of establishing the interfaces. (To amplify what I mean by "personalization," I have streamlined the interface procedure and adapted it to general requirements. That has produced the 12 block diagrams in Figs. 12-5 to 12-16. Though the diagrams refer to the foreign trade files of a bank, it is understood that the procedure is generally applicable to the handling of any masterfile.)

The processing goal of the proposed block diagrams is as follows:

1. Data selection (extraction)

2. Sorting in preestablished order

3. Parameter-driven programs able to convert file structures to one homogeneous basis

4. Data validation, assurance of conformity to ranges, and insertion of color code for exceptions

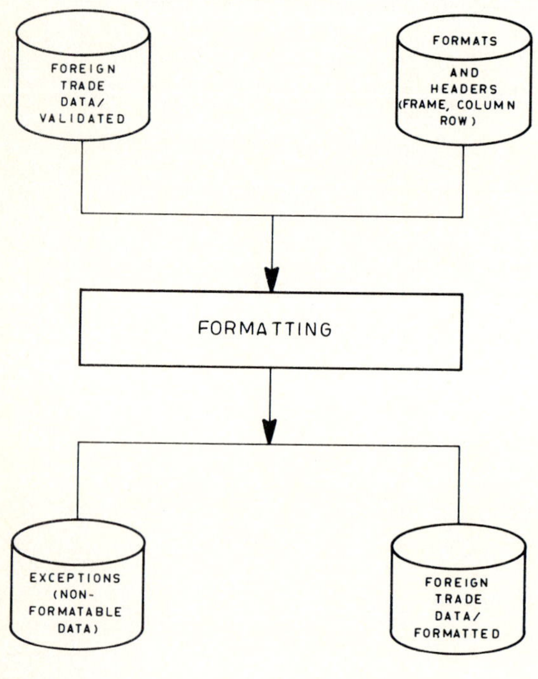

Figure 12-9 Format into established structures.

5. Formatting into established frame structures

6. Color flip-flop for easy reading

7. Creation and maintenance of historical file

8. Graph preparation

9. Formatting of graphs to VDB protocol; code insertion

10. Security codes, exclusives, and protection procedures

11. Consolidation of text, data and image frames; filing of tree structure

12. Send routine; image recovery

Particular attention must be paid to the handling of parameters and the application of the necessary controls. In the Whitbread application, all files created by the system have a trailer record on which is kept the count of records in the file. Each program reading a file created by the system checks the number of records against the count held on the trailer to identify discrepancies. Details of the reconciliation error are

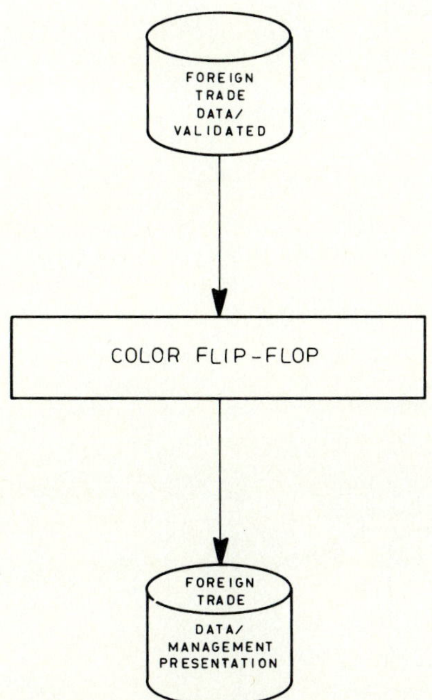

Figure 12-10 Color flip-flop for easy reading.

printed on the appropriate record. For restart and recovery purposes, files are copied and retained for a period in case reruns are necessary.

Cluster security is assured from the origin of the information: A cluster identifies a whole handling procedure through the terminals allocated to an application. This is a very important concept, and it can be applied to a closed user group.

As we will see in a subsequent chapter, another British user assigns color to indicate the origin of the reports: blue for personnel, red for finance, and so on. I would not advise a full-color page for this purpose, but a leading line at the top of the page (preferably the second row, after the header) can advantageously be assigned a color that indicates origin and can be shown consistently with that identifier.

Cluster security, the assignment of identifiers, and the use of trap codes must be examined in a consistent way in the design phase. Other security measures to be taken have to do with updating. Related to them is the problem of making a selective updating capability feasible. For instance, a program will not extract information from a file whose

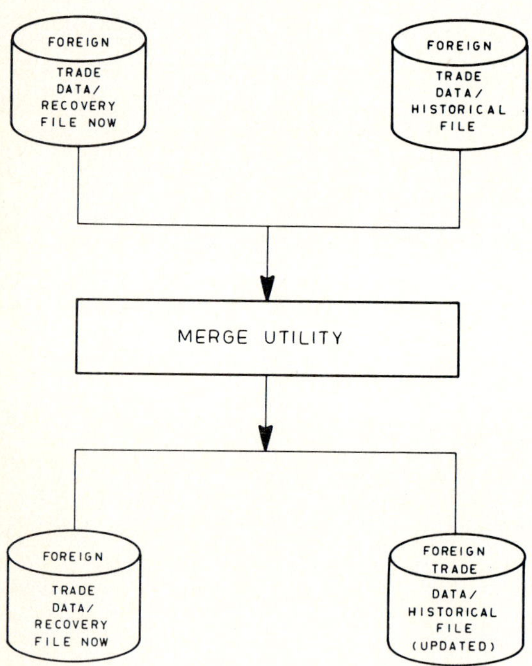

Figure 12-11 Creation and maintenance of a historical file.

control dates are the same as those held on the "cluster security file" for that same cluster. Normally in this application that indicates that a new file has not been loaded for that cluster, but the controls which have been put in place by the systems analysts lend themselves to several applications.

In my judgment, many computer-based procedures are wanting only in regard to the share to be given to paper output for control and other purposes, such as validation reports, input lists, and error reports. These should definitely be replaced by online data capture and validation with a quality database use if necessary. Rather than print backup reports for use if the videotex equipment is out of order, I recommend COM or videodisk output. It is, however, wise to make the backup output match the image of the videoforms. This definitely simplifies the training of the end user and makes the system more user-friendly.

Another excellent feature of the new systems used for decision support is the clearly established relation of videotex presentation to existing batch-oriented subsystems: whereas for top management information the degrees of freedom in designing the application may be many,

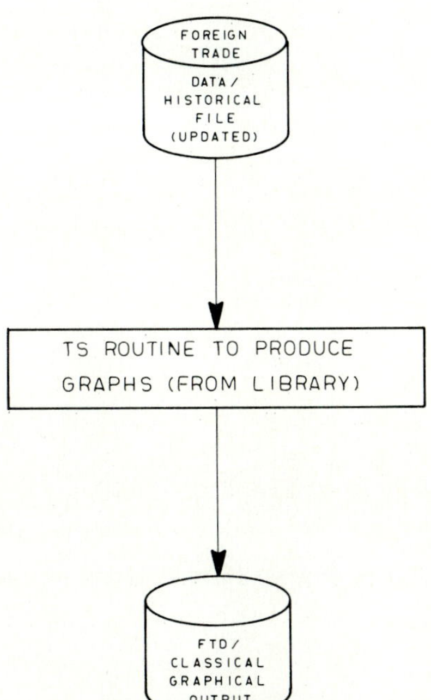

Figure 12-12 Graph preparation.

for middle management there are already available a number of computer runs which have conditioned the way responsible executives look at data processing.

Hence, the procedures established in elaborating a relation with existing subsystems had to do not only with the manner of presentation but also with the timeliness of the information being presented and other issues. An example of the latter in the Whitbread application is that the cutoff points, which with the batch solution tended to vary from depot to depot, have been consolidated. Furthermore, particular care must be taken during the planning period in establishing the nature and extent of the interactive phases assigned to management and of the procedures to be followed in selecting from bulk data. An interesting aspect of this subject is the elaboration of groupwide units of measurement in accordance with the end user.

Cross-indexing, made possible through videotex, has also presented design challenges. It was not available with batch processing and was

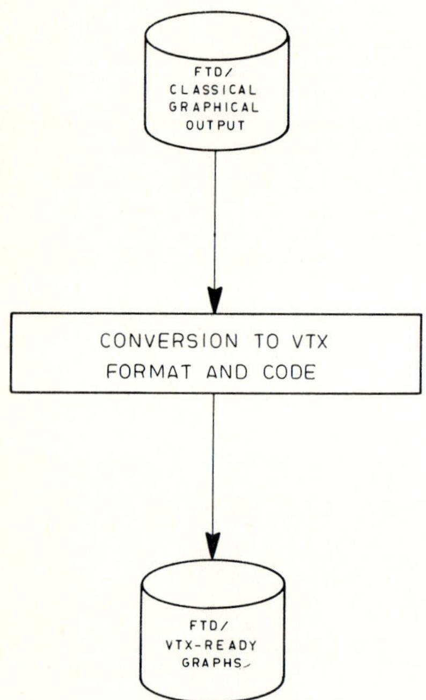

Figure 12-13 Formatting of graphs to VDB protocol and code insertion.

therefore unknown at the user's end prior to the introduction of video-tex capabilities. To help the user learn the system, the systems analysts should pay proper attention to *prompting,* another facility available with videotex but not feasible with batch.

An easy example of prompting is at the editing level. By entering the edit option, the editor will, most likely:

- Amend the existing frame
- Copy the existing frame
- Delete the existing frame
- Enter a new frame
- Interrogate frame activity
- Overwrite the existing frame
- Free modification status

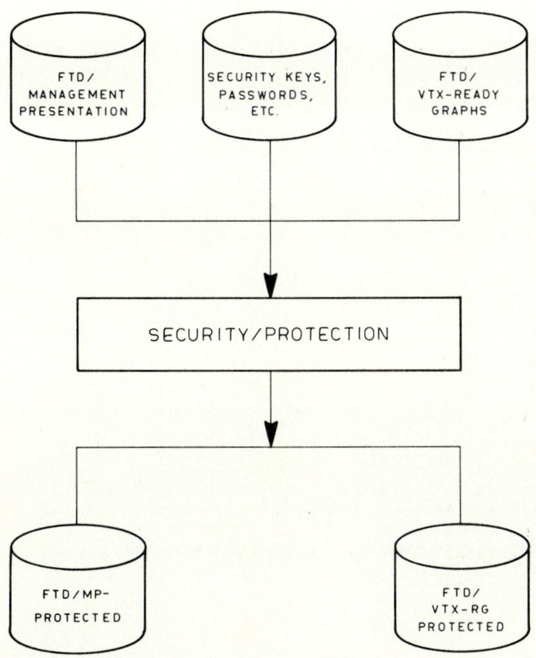

Figure 12-14 Security codes, exclusives, and protection procedures (passwords, traps, and transparencies).

The videotex page should give the user those options together with the choice of switching to the proper routine through one key stroke.

Prompting means that several pages of helpful information are available to the system user. To access any page all the user needs to enter is the number identifying the subject he wishes to explore:

- Broadcast
- Monitor
- File maintenance
- System control
- Keyboard layout
- Port status

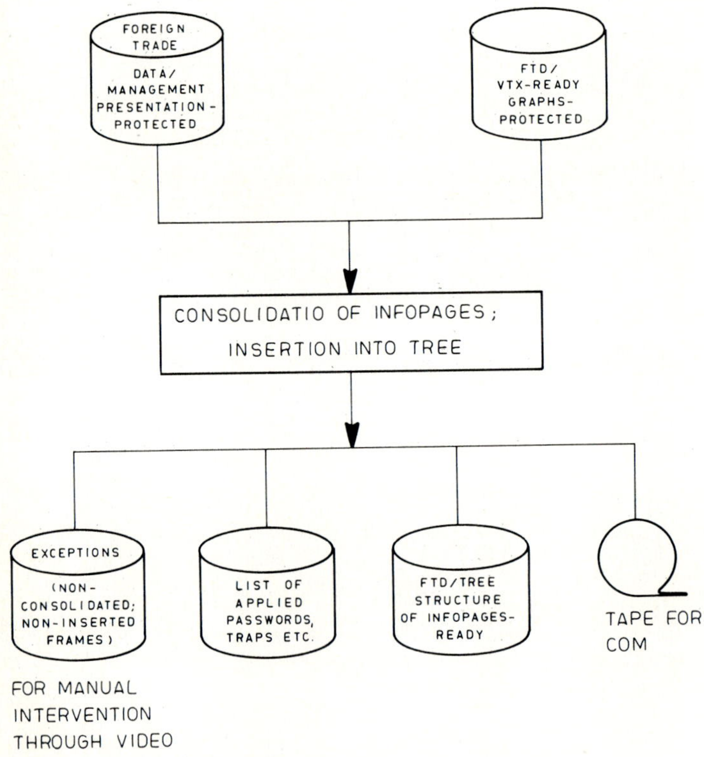

Figure 12-15 Consolidation of text, data, and image frames and the filling of the tree structure. The alternatives are the daily (selected frames) procedure and the full procedure.

And he can return to a page through a simple key stroke. Table 12-1 presents an example from an application in which the system eases the signing up of the user through the process of registration. For prompting purposes, it is wise to devote one or more pages to an executive as note pad and organize these pages in a self-tutorial way.

In another case, reference pages have been added to handle the operating company's index and enable the user to locate his company's (divisional) pages. Content, for instance, involves:

- Operating company description
- Depot page reference
- Product page reference
- Operating company page description

Such references point directly to the beginning of the summary screens for the company. Other reference pages include product description,

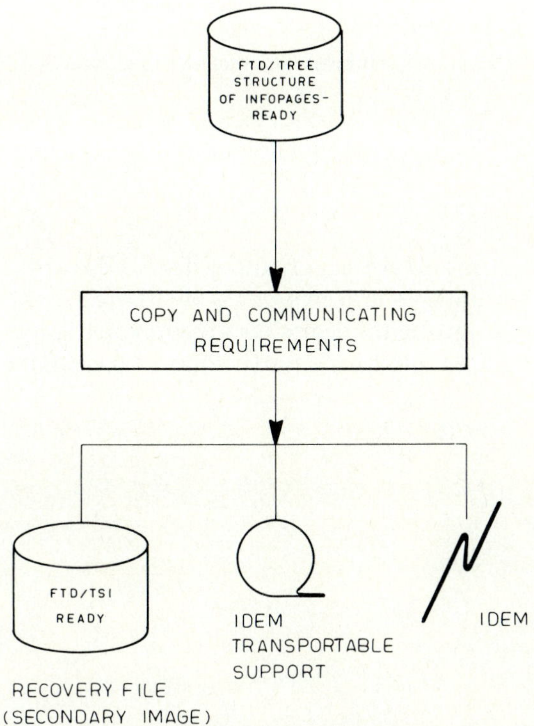

Figure 12-16 Send routine and recovery.

Table 12-1   Helping the User Sign Up through System-Supported Prompting

Questions	Answers							
	1	2	3	4	5	6	7	8
Does set transmit own ID?	Y						N	Y
Is set registered on system?							N	Y
Do you have registered password?							N	Y
Do you require mailbox?		Y					Y	N
Do you require editor?			Y				Y	N
Is the page required in the user group?				Y				N
Are you a member of that user group?						N		Y
Is "frame" access level greater than yours?							Y	N

1. See guide supplied with set.

2. To use mailbox for receiving messages, you must be registered for mailbox with a user's short name. To send messages, you require the above plus an edit keyboard.

3. To use editor (write information into system) you require an edit keyboard and you must be registered for editor.

4. You must also be a member of that user group.

5. If frame or page belongs to a user group, you must be a member to gain access.

6. Access will be denied.

7. Contact DBA for further information.

8. Good. Continue to next question.

product sequence, receipts, due date, days' cover, and different days and levels of delivery. Among other notable improvements over batch is the rounding up of details to three significant digits for videotex presentation. That is exactly what management needs for the operations under its control.

The Mechanics
of Interfacing

*Success depends on knowing how long it
takes to succeed.* MANAGEMENT PROVERB

Among the problems which have not yet been resolved are the mechanics of interfacing between the corporate resources—text and data stored on mainframes and minis — and the VDB which will be accessed by the users. A long, hard look at this subject identifies the need for serving a population of three types in an efficient manner:

1. End users (consumers, company executives, clerks), who retrieve and employ the information in the VDB

2. Information providers, who create and maintain the text and data in the infopages

3. The system manager, who controls user access, realtime update, bulk update, and associated accounting data

Each class of users must have available features and operational modes which permit efficient action. Not only should the retrieval system allow users to choose and display frames of information but also the relation of the VDB to the greater store of information handled by the organization on big, bulky, and expensive mainframes must be properly established.

On numerous occasions we have emphasized that videotex systems, in order to become widespread, should be simple to use, cost-effective, and attractive in terms of the information they provide. Yet, although at the consumer level a telephone line and TV set may suffice for the end user to reach any database anywhere in North America, careful preparatory work is necessary at the information provision level. This is effectively the message which was conveyed by Chap. 12 and will be completed in the present discussion.

We have spoken of extracting and editing. Part and parcel of those functions is indexing if we wish an index inquiry facility to be used to list a specified set of frames from the VDB. That operation, too, has its prerequisites. Before producing an index, we must specify:

- The frames within the VDB to be examined
- The criteria which the frames must meet

System software will then be called into use to display an index which may run to several screens. Three pieces of information are needed: the frame specification, the selection criteria, and the listing level. The frame specification determines which frames in the database are to be examined when the index is produced. There are basically two options in obtaining such software: buy it or write it. Purchase is the preferable solution; but at times it may be necessary to write the software, as the following example helps demonstrate.

Project Interview

Project Interview was set up as a cooperative effort by three organizations after they failed to find a suitable package for the conversion of daily mainframe-based data to infopages for videotex utilization. Individualization, codification, and security of information were primary objectives.

The designers paid particular attention to the preparatory work in the control and formatting phases. They put due emphasis on the classification of input file parameters and record parameters, as well as the definition of tables (arguments and functions), text screenfuls, and other outputs.

The parametric approach to formal design and control procedures is a fundamental requirement for automated data extraction and interactive presentation, which permit the transfer of information produced by existent data processing tools (typically structured in files fitting those tools) to the VDB. To clarify that statement, let us take the procedural approach and consider the preliminary activities which have to be developed for data presentation through videotex.

The basic supposition is that the IP has individualized the information for the videotex system, which means that he has defined his "arguments." An argument helps individualize the wanted text and data, the chosen mode of representation, and the outlining elements on the screenfuls such as:

- Background color
- Foreground color
- Textual headings (infopages, columns, rows)
- The presentation of text and data

Interview handles those elements, which, however, must be specified by the IP with a four-digit code (from 0000 to 9999). If the same data are represented in both graphic and alphanumeric forms, two different codes are assigned to each form of representation, assuming that the procedure calls for two different arguments.

Selection criteria help determine which of the frames specified will actually be indexed. The user may select all frames, frames created or altered today, or frames created or altered but not yet sent via bulk update. We will return to this issue. The listing level determines how much information will be included in the index about each frame. The user may choose just a count of the number of frames satisfying the selection criteria, the infopage numbers, routing information, access counts, date last altered, and so on.

The next basic question is the creation of routing pages and title frames (multipage descriptors). This is an organizational activity. In the Interview procedure tree structures are not created as such; the respective frames should be inserted through a terminal. Similarly, elements relative to the addressing of the infopages (frames) should be specified by the IP prior to the assignment of the associated indexes.

To meet these difficult prerequisites at the end user level, the editor routine offers not only the standard features contained in any videotex editor but also extended facilities to make the job of creating and maintaining frames easier. Three main modes in the editing subsystem meet this requirement:

- The frame control mode lets the administrative options and attributes of a frame be altered: create or delete a frame, change the frame text and/or frame control information, or duplicate a frame.

- Through the frame text mode, the contents of the specified frame are displayed on the video screen. If a new frame is created, a blank screen is shown, except that the first line contains text selected from the frame control information which is present on every frame.

By means of this facility, characters can be entered at a position and overwrite any existing text; graphics, color, flashing, double height, background, and separated graphics can be employed; the cursor can be positioned by using the cursor movement key; response fields can be created on response frames; and so on.

- In the extended edit mode the bottom two lines on the frame are used to handle selected options. For instance, the appropriate color code must be activated, and then the information required for the editor to carry out the operation will be requested.

To assure confidentiality, the IP must assign to each argument the appropriate code (trap). This reflects the level of authorization management has chosen for each information element. Regarding the management of the file parameters, Interview presupposes the creation and upkeep of a parameter file containing all directives and aiming to:

1. Individualize the input data to the VDB (which must already exist on a magnetic support on the mainframes). The basic operations of such a program are to create, store, get, change, and format files that are needed by the videotex systems.

2. Define background and foreground color and the layout of the text on the videotex screen (including fixed headings and options).

3. Define the form of representation (color, flash, etc.) of the data on the infopage.

The data extracted from the mainframe (or mini) storage medium and used as an input to the VDB will be entered in the parameters file run by the system. If necessary, input to the parameters file can be manipulated through a video terminal on the mainframe (or mini). In this particular case the Interview procedure is to utilize conventional file structures and accept as an input one file at a time from mainframe storage. There is substantial flexibility in the structure of that file. The identification element of a file is reflected in the parameters of the formatting program.

An orderly job requires that every record in the file find itself in a properly determined position, given that its coordinates act as parameters. For the conversion of tabular information into graphical form a subroutine written specifically for that purpose is used. Within a given label, different data fields can be identified. In the Interview program one file is handled at a time; each field to be individually manipulated within the file must be properly labeled. The user is free to employ code names of his choice except for the word "data," which is interpreted by the program as a visualization command.

If a manual procedure intervenes before digitization is begun, it is advisable to design the screenfuls on a module. All codes and descriptive elements which serve as definers should be noted:

- Background color
- Captions for columns and texts
- Format of data to be edited in output
- Foreground color, flash, and so on.

Numbering rows and columns facilitates the individualization of the coordinates of each information element. Since Interview has been optimized within the context of the British Viewdata specifications, each command for color, flashing, double height, and so on uses one character. The program itself takes care of the command characters which it assigns, but the person intervening through an editing terminal to effect changes must be careful not to overlay already assigned characters—unless he specifically wishes to do so.

The same observation is valid in the sense that a double-height heading occupies two lines and care must be taken to avoid overlaps. Also important is the execution of formal controls on the parameters. such controls are effected automatically by the program in the structuring of an infopage and by the editing terminal in the case of a manual entry. The program also provides linkages between the parameters employed in different video screens in order to assure the presence of all the elements necessary for formatting.

Utilities in the Interview library look after administrative operations in the preparation of the input to the VDB such as:

- Query
- Insertion of data
- Deletion of data
- Other types of modification (color, size)
- Editing out of whole infopages

Similarly, in terms of feedback, the number of accesses of each frame in the VDB is recorded for statistical calculation purposes. This information is stored each time a frame is updated, and it is kept in the same record as the frame text. In addition to the record use on a per frame basis, user accounting information is maintained. The following data can optionally be displayed at sign-off time: page access to date, page access this session, cumulative time used to date (hours and minutes), and time used this session (hours and minutes).

Performance measurements enable the system manager to monitor and control system operation and the service provided to the user community. This information is logged to disk, and a standard utility program is provided to generate reports covering average response time as a function of disk access traffic and frame access traffic, response time distribution, average input and output loads, and response time as a function of input and output loads.

Keeping track of uptime, response time, and other operational factors is a help in spotting and heading off trouble before it gets out of hand. Monitoring user attitudes helps focus the videotex service on user needs. But although measuring operational parameters is fairly routine, we have to take special steps to find out what users are thinking.

We should monitor projects that are particularly vulnerable to foulups as a means of knowing promptly when a poor situation begins to sour. In that way we can keep alert to potential difficulties while we are setting the recovery in motion. Vigilance is important. Our image as competent professionals suffers considerably less damage when we discover our own mistakes than when we are taken by surprise. We can limit the fallout from projects that do not fly by not raising expectations about them higher than we have to and by following them up in the right manner.

Software Routines

We all know by experience that the key to exploiting the potential offered by a computer is software. This is a valid statement whether we talk of mainframes, minis, PCs, or videotex systems. The following is a list of utilities which can help enhance the service the user can obtain from his investment, starting with what the British call the Viewdata plus solution. The term was originally employed to identify the possibility of adding data entry to videotex as a convenient way to specify and update the database. Typically, data entry frames are blank pages to be filled by the user and transferred by the system to the mechanism which handles the database, whether corporate, regional, or functional.

Another integral part of a utilities library is the text editor. It puts the installation firmly in control of logic, data, and reporting formats, assigns index numbers to provide quick access to anyplace in the file for subsequent insertions or changes, and contains routines for fixing, duplicating, and moving lines around.

A third key routine is the report generator. It interprets a report specification file and takes numbers from the work area to produce screen displays, printed reports, or files containing reports. In addition, it can be used to scan a logic file and print out a data input worksheet for all input variables in a model. The report generator gives control over which columns and rows to include in infopages, print out, and in what order. The editor can use row names from a logic file, specify different ones, or use no row numbers.

Typically, a report generator will have complete control over the number of decimal places in each row and column. It will also control the width of each column individually, insert percent signs after numbers and dollar signs before numbers, assure single or double underlining and the centering of titles, look after the suppression of rows with all zeros, provide brackets (or red color) for negative numbers, and prompt for comments.

A fourth routine, the graphic generator, will produce color charts, line graphs, and pie charts from numbers contained in the work area or any specified file to be displayed through infopages. As an example, an industrial firm developed routines to support through videotex the following types of graphical reports:

1. Bar chart of net profit margin

2. Bar chart of times interest earned

3. Pie chart showing how sales dollars are spent

4. Bar chart of fixed and variable expenses month by month

5. Line graph of sales, gross profit, and operating expenses

6. Exception reporting through color identification

When we display text and data in different colors, we use the colors to convey information. We can highlight certain data, call attention to a particular piece of information, and group certain data to make it easier for the user to see it. We should look upon color not as a way to make the displays pretty, but as a way to add another dimension to the information that we can display.

Similarly, a specially developed routine looks after consolidation: the combining of results from references representing similar entities such as company divisions, geographic regions, product lines, and so on. After consolidation, the routine generates tabular reports or graphics, provides details or highlights on command, and sees to it that the reports or graphics contain the information in the most desirable format.

These editing routines are particularly oriented toward executive information systems. By utilizing color graphics, we can cut through details to show the relations and trends that can too often be obscured in numeric form. Volumes of normally incomprehensible statistics can be quickly and easily browsed through to target in on areas that require attention.

Color graphics employing exception and variance reporting techniques allow us to see at a glance when there is some change in a trend or when the data departs from expectations. Interaction with computer-based graphics information gives us the ability to summarize quickly massive quantities of data and translate voluminous numbers into meaningful, action-oriented information. Not only do graphs make it possible to show more data, but the job can be done with less paper and be much less time-consuming to the executive than classical reporting techniques.

Graphic presentation is especially useful for pointing out trends over a time period. We can present time series data for different items in the same graph and still get the message across, which would be impossible in a tabular form of graph. The multiple-group plots are particularly popular because they can show absolute quantities on the top part of the graph and proportional relations on the bottom.

It is the old 80/20 principle applied through videotex: We make 80 percent of our decisions with 20 percent of the data, and we can show data and graphs online and interactively through the system. Editing, formatting, and reporting through a videotex software mechanism:

- Bypass the rather substantial cost of computing a chart on the mainframe and the use that each new mainframe-connected terminal implies

- Cancel the additional operator time needed to run a mainframe system

- Avoid degradation of other mainframe applications that are trying to run at the same time as the graphics program

- Underline the value of local data and picture storage that is independent of mainframe or mini support

- Override the speed considerations imposed by the low-speed link between the terminal and computer

- Assure the manager a quick response with extremely simple system operation

We must, however, guarantee communication between the database and the mainframe (or mini), and to do so we need to transfer information between the videotex engine and the host. For that there are two possibilities: bulk transfer (which we will examine in the next section) and an interactive transfer of information that permits a transmission rate commensurate with the job on hand. Typically, this program is menu-controlled, and prompts are given to the user so that he can decide on his next operation.

A Bulk Editor Program

Bulk update is a term used for the method whereby one computer (typically mainframe or mini) in a videotex system can update another. This could be a public system or another private machine. In either case, the bulk editor should be capable of operating in both environments, thus both generating and accepting bulk updates conforming to the agreed standards.

Ports on the system can be configured for user terminal access, bulk update input, or bulk update output, but they can not be mixed. Through a bulk update input port, the system must respond to the incoming requests and updates in the standard way. For an output port it is often necessary for the user to specify the update activity to be performed.

The *bulk transfer* function is concerned with the bulk transfer of pages of data from any supported database to the VDB. While that function is being performed, the block sequence numbers are checked to ensure that no data is lost. This is the only time that sequence numbers need to be checked. Let us now look into the operations in greater detail.

Once the management-oriented frames have been extracted from the general database, they will have to be loaded on the dedicated videotex machine. Loading can be accomplished either online—mainframe to videotex engine—or through a magnetic support medium (cartridge, disk, or tape). The former solution is preferable, but for daily availability purposes, in case there is a problem with the lines, both alternatives should be worked out.

The use of a bulk editor program requires one, as a prerequisite, to first work through the extract program and then format the videotex frames. The output could, for instance, be a tape to be written in a form specified by the bulk editor. This tape must be brought or transmitted to the videotex machine, then the bulk editor will perform some functions (insert dates, and so on) and load into memory. For up to 3000 frames, one such system seems to work well.

Figure 13-1 identifies the work to be done in order to provide the interfacing between the mainframe and the VDB:

- Extraction is applications-dependent

- Formatting and loading are general-purpose

While this work is being done the future capabilities to be added to the basic software with a new release should definitely be kept in mind. Examples are data collection, store and forward, and main index reference. The latter means that, rather than go through the top of the database pyramid it is possible to follow a special, main index subdatabase.

Development work on this or any other system should take into account the fact that a package has a limited lifetime. Every year or so, a manufacturer comes up with a release which supports more functions than the one preceding it. Figure 13-2 demonstrates the interface software to be developed on two levels:

1. Converting existing program files from the spool for printing to a stack of videotex frames in the making (or from the master file itself)

2. Writing new programs able to do the necessary extracting and formatting jobs

Ideally, the second should be parametric, support user exits, and be open to expansion as new facilities develop and/or are required. All information about color, graphics, flashing characters, double heights,

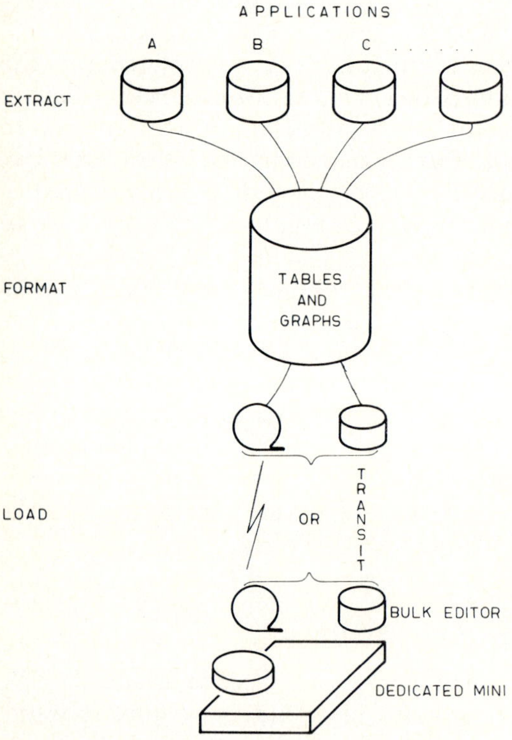

Figure 13-1 Providing the interface between the files (on mainframe or mini) and the viewdatabase is one of the fundamental jobs to be done in setting up the system.

and different backgrounds must be inserted at the editing routine level. That is the user's job (more precisely, the job of the user's own specialists), and it must be done well.

As new features pile up, it is necessary to change generations. This means that, by 1990, the videotex system is in for a major overhaul. There are technical reasons also. A high-resolution TV system is one example of why basic changes will be necessary, and the nanocomputer (pocket computer) is another.

Finally, an integral and very important part of the study which needs to be made is that of the systems support to end user functions. Typically, it should involve:

1. Starting with the choice of the application

2. Studying the videograms with the end user

3. Collaborating with the specialists responsible for the interface pro-

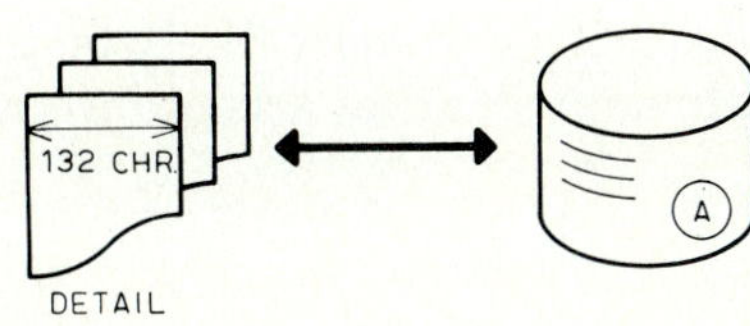

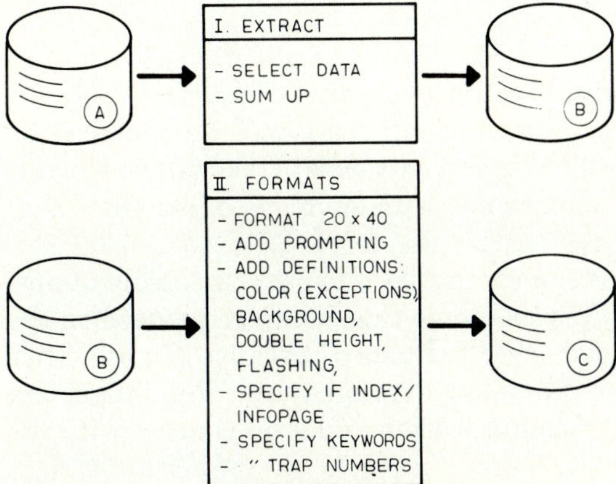

Figure 13-2 Interface software can be developed at two levels: converting existing files from a printing spool and/or extracting and formatting data from the updated master files.

grams for organization, mode of presentation, color, graphics, and exceptions

4. Working with the DBA on database organization and also authorization and authentication

5. Providing assistance to the software experts on the design of prompting and explanation frames

6. Helping the end user on babysitting, queries, and time out

7. Assuring EUF-type maintenance

These activities are no "strange requirement" of videotex services. They are part and parcel of the applications support to be given to any modern information system, and they become more important as we move into a management information systems environment.

Being User-Friendly

With videotex, the key words are simplicity and an impressive market perspective. It is widely acknowledged that, as far as new office tools are concerned, the biggest market of the 1980s will most likely be the *text* and *graphics* data terminating equipment. Market acceptance will be greater if solutions present:

1. Minimum programming cost

2. Ease of use

3. Good cost reduction perspectives

4. Highly structured capabilities

5. Preformatted approaches

Videotex and facsimile terminals can provide those features. As far as the user layer is concerned, programming costs will be replaced by *consumable telesoftware,* another reason why ease of use will be a prime factor. Anyone who has had experience with such systems makes that statement and sticks to it.

The *user-friendly protocol* is something the end user can easily understand and follow. The protocol is so simple that it can be printed on the keypad. This has been a good BPO influence, because a post office organization must design for the below-average man. The danger enters when one tries to increase sophistication and loses contact with the user.

The TV set idea kills two birds with one well-placed stone. Its psychological advantage is that we know how to turn the TV on and off and play with the control keypad. But also the cost is low. It is only a fraction of the price commanded by a specialized computer terminal, and it also supports color and graphics. Savings such as that explain why, as Fig. 13-3 suggests, videotex and facsimile present the possibility of steady cost reduction throughout this decade to a level well below that of the more intelligent communicating processors.

Costs aside, the psychological factor is sure to have a tremendous impact on the use of this interactive service. To have said "Here it is. We now have a new computer solution" would have had little effect on the average user because the average user is afraid of computers. But we do not say that. What we say is: "Use your TV set."

Improvements are, of course, in the works, but they should not involve the *end user*. That statement is not exclusively directed at the average man, the final consumer. Even companies with considerable expertise in information systems have had their share of problems with realtime, timesharing, MIS, and other expensive arrangements which did not work as expected and overran their budgets.

Since its original release in England in 1979, Viewdata has supported a cost-effective approach to selling expertise to people who want to communicate with the public. But its future as a new office tool will be better appreciated if we keep in mind that it evolved quite rapidly beyond the Prestel standards to include such services as Frame Freeze and Videopaint. Meanwhile, its inherent limits provided insurance

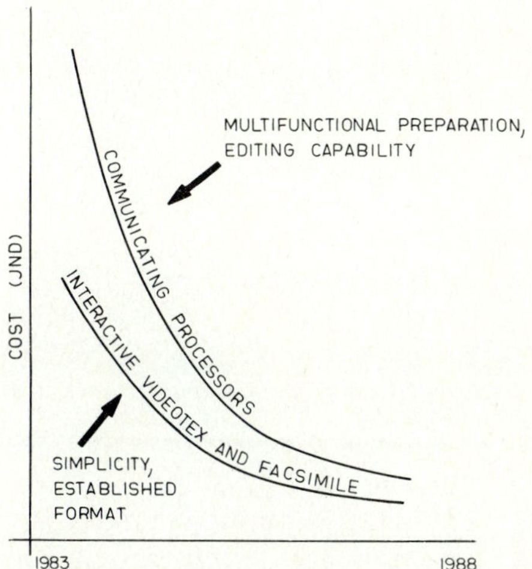

Figure 13-3 Characteristic cost curves for videotex, facsimile, and communicating processes. The potential for important cost reduction is present in all alternatives.

that there is no probability it will skid into uncontrollable sophistication.

The benefits are twofold: First, the typical user can work without changing in any appreciable way his frame of reference. The tool is easily adaptable to office chores; its use does not involve the appreciable changes caused by more flexible, computer-based systems. Second, we benefit from advances in technology and preserve the capital investments in software at the same time. As Fig. 13-4 demonstrates, because of advances in the physical sciences and engineering design, hardware capabilities in MIPS (millions of instructions per second) make big strides every 2 or 3 years. Hardware is no longer the limitation, but rewriting the aging software is a major problem—and one that we can avoid.

The price is discipline. We have spoken of the need to support structured capabilities and preformatted approaches. Videotex does that both in terms of the dialog procedures and in regard to the *response frames*. And let us not forget that ease of use, low cost, and procedural structure (hence, user discipline) come hand in hand.

With each videotex set for office or domestic use comes a small remote control and an arithmetic keyboard similar in appearance to a pocket calculator. With the keyboard, interactive users call up frames from the VDB *as if* they were changing channels or varying brightness and

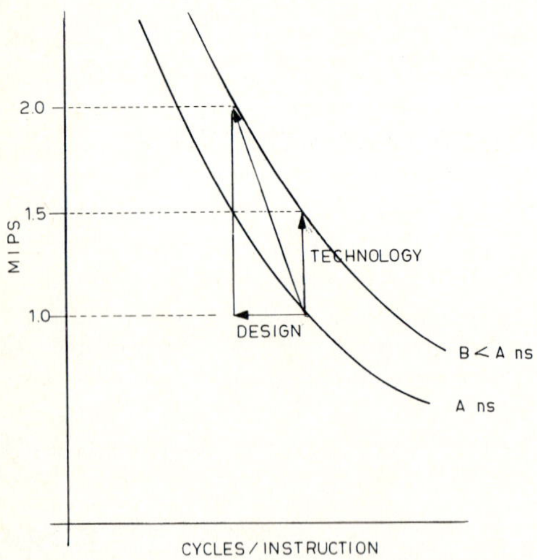

Figure 13-4 Whether through advances in the physical sciences or in engineering design, technology sees to it that higher MIPS are obtained at lower cost.

volume. Information (pages) is retrieved by pressing numbered buttons on the unit. On business sets, this control could be replaced by an alphanumeric keyboard with or without function keys or, even better, a personal computer.

To get Viewdata information, English users first switch on the TV set and then call up their local Prestel center by pressing a button on the control. They do not even have to lift the telephone receiver. Then, at another touch of the button the Prestel opening display appears on the screen; it is an index listing the main headings under which information is available.

By following a simple guide displayed on each page, a user follows a route into the VDB (which we will explain in detail in subsequent chapters) to find the particular page wanted by pressing buttons on the control unit. This is a *menu selection* approach. Alternatively, a viewer may remember the page number or can look up the number in a printed directory, and select the page *directly* by pressing the appropriate buttons on the control. Three directories have already been produced by the BPO for home users and supplied free, and there is also a business user guide.

The instructions to the Viewdata user are as simple as ABC:

- Switch on your television set.

- Make sure no one is using the telephone.

- Switch to Prestel according to the operating instructions supplied with your set.

> After a few seconds the Prestel page welcoming you should appear on your screen. The computers in the network provide for closed user groups (CUG) which enhance the privacy and security of the Viewdatabase. In different terms, Prestel may refuse to give you access to certain pages. And the same is true of the in-house systems. This is because these pages belong to a CUG. Its members pay a fee to the information provider who owns the CUG in order to be permitted access to these pages.
>
> Interactivity between the ultimate consumer and the Viewdatabase is assured by means of the response frames. These are a special kind of Prestel page where you give a message to the information provider who owns the response frame. Response frames contain a space for the user to write on the screen using the Prestel keypad. His message is sent to the information provider via the Prestel computer.

For receiving screenfuls (Prestel pages), the reception rate gives a buildup time of 3 s for an average page, and with duplex working the user can interrupt page buildup if desired. The frame format—24 lines of 40 characters—is compatible with teletext. The system provides for flashing symbols as well as simple graphics and eight colors (including black and white).

The network has been designed to keep transmission distances short, thus permitting the modem within the TV set to be kept simple and comparatively cheap. This means that virtually all customers in the local fee and adjacent call charge areas should be able to obtain local Prestel service without difficulty. Outside those areas, long-haul connections are possible, but each connection must be examined individually to determine line loss, noise, and other factors in degrading transmission.

There is good evidence why prospective users should carefully avoid rewriting the software. The available package has commendable functions, and it is better to improve on it. We must account for the fact that we are actually entering a new era of data processing. The potential market will be enhanced through standardization and the ability to handle a universal system. Standardization also means lower costs.

Furthermore, to make the market grow, it is important to be able to sell the device as a tax-deductible business expense. That is what is happening now in England. Businesses which subscribed to Prestel give their executives terminals to make it feasible for them to communicate with their offices through the Viewdata service in their living rooms. And it will most likely turn out that the business market is the best way to get into the domestic market and expand it. The energy crisis will help in this transition, because mode of travel of business people going to work is coming more and more under scrutiny. And the further-out thought has occurred to some executives: Videotex can allow 2 days of work in the office and the balance at home, like Swiss watchmakers of the old days. That is to be seen.

Telesoftware

Telesoftware means remotely loaded software to run on a minicomputer or intelligent terminal. Programs can be stored as pages of information in the database of a public utility or company-operated and/-maintained network (COAM); the user can remotely select the programs he requires. To a substantial extent, telesoftware is consumable software.

As the pages are transmitted to a user's personal computer, they are charged, thereby supplying the program provider with revenue for his product. There are problems to be solved in this connection:

1. Portability

2. Error probability

3. Billing

Portability can be assured through networkwide standards (which is unlikely) or by bypassing virtual terminals and providing the interfaces.

A prerequisite of telesoftware is a programming language which is truly portable so the same program code can be run on different microprocessors with the appropriate interpreter. A software company, CAP-CPP, has already written interpreters for the most common teleprocessors and claims that it is a simple task to produce an interpreter for any microcomputer. For instance, MicroCobol programs have been loaded on the Prestel database and can be accessed by intelligent terminals with CAP's Viewdata terminal programming (VTP) software. The user accesses the Prestel database, and then VTP takes over and loads the MicroCobol program into the terminal's memory.

Error probability is controllable through error detection and correction, the same methods we use with data. It is also possible to provide for standard local tests at the recipient's level. Billing is straightforward. Each time a telesoftware program is retrieved, the program provider receives the appropriate frame access charges.

The challenge with billing is evasion and abuses. Two answers are projected:

- *Price is the protection.* If it costs less to purchase than to reproduce, the user will purchase.

- *Automatic destruction after, say, five uses.* For instance, one program provider has built time lockout switches which allow the programs to be used for a predetermined number of days after they have been retrieved. Thus, an infrequent user will have to pay each time he wants to use the program. This is something we are going to hear more about in the coming years. It has a correspondence in other fields, such as the microprocessor-controlled credit card.

Receivers with a telesoftware capability are expected to appear on the British residential market. The typical device will consist of a TV screen, microprocessor, alphanumeric keyboard, and possibly storage in the form of diskette.

A disadvantage of telesoftware is the time taken to load a program into the receiver. A program of 2 kB (kilobytes) can be stored in five Viewdata pages. A page takes 8 s to transmit. Another 4 s is required for VTP to decode the transmitted data and load it into the microprocessor. Hence, programs can be loaded at a rate of five pages, or 2 kB, per minute. But only a very simple program can be held to 2 kB;12-kB routines take 6 min to load from the Prestel database into the receiver.

The time delay issue can be solved nicely through broadcasting, because telesoftware can be transmitted via TV stations at 250 pages/min. The BBC experimented with software transmission by Ceefax. But broadcast teletext is not interactive, and no charge is made for the information received. Thus the answer will most likely be terminal enhancement and parallel handling. That will require local intelligence through microelectronics. It is easier to build a personal computer than a TV set, and personal computers can integrate nicely with videotex networks.

Gaining Insight

*If a research project is not worth doing at all,
it is not worth doing well.*   GORDON'S LAW

To paraphrase Gordon's law, if a project is worth doing, it is worth doing
well. To do a system job well, particularly on the frontier of applications
knowledge, we must gain significant insight. In that sense, there is no
better teacher than experience: We have to talk to the people who
designed and implemented videotex systems.

Software for videotex is available on microcomputers, minicomputers, and mainframes. The user has the option of buying a readymade
package including software and hardware—which is most advisable—
or writing the software himself. Dr. P. S. Arbouw (of IBM Netherlands)—a graduate student of mine in a seminar on databases offered
in Berlin (March 1981) and sponsored by George Washington University—wrote a simple Viewdata routine in about an hour. It is shown in
Table 14-1. The writing in standard IBM macros (excluding the page
management) requires 33 statements.

- 23 to rule the display capabilities

- 10 to rule the account information

The message and answer capability in the system (for example, subscription to lectures) was provided by the standard system software.
This exhibit is an abstract of routines originally written in VM/CMS
Exec Language (the statements mentioned in table 14-1 in another
format) by using IBM's VM/CMS library. The latter contains all the
display pages, the response files, and the accounting file.

The display input-output facility (DIOF/CMS) was employed for displaying the pages, storing messages, dealing with the answers, and so
on. The original version was in four colors, and the objective was to

Table 14-1   Base Code for a Viewdata-like System

Legenda:	*	THIS STATEMENT IS A COMMENT
		THIS STATEMENT IS ONLY FOR ACCOUNTING REASONS
		(BETWEEN BRACKETS IS A SYSTEMS-DEPEN-DENT BUILT-IN)

```
INITIALIZATION    *    THESE STATEMENTS ARE FOR CHARGING
                       STARTTIME = SYSTEMTIME .
                       PAGECOUNTER = 0 .
                       FEE = 0 .

START             *    INITIAL PAGE AT START AND RESET
                  CURRENTPAGE = 0 .
                  PREVIOUSPAGE = X .

DISPLAY           *    DISPLAY REQUESTED PAGE AND ACCOUNT-
                       INFO
                  *    TEST EXISTENCE IN LIBRARY
                  STATE CURRENTPAGE .
                  IF NOT FOUND GOTO NOPAGE .
                  DISPLAY CURRENTPAGE .
                       PAGECOUNTER = PAGECOUNTER + 1 .
                       IF RESHOW = 0 THEN
                               FEE = FEE + PAGEPRICE .
                       RESHOW = 0 .

                  *    TEST THE GIVEN ANSWER
                  IF ANSWER = * GOTO BACK .
                  IF ANSWER = 9 GOTO START .
                  PREVIOUSPAGE = CURRENTPAGE
                  *    NOW WE HAVE AN ANSWER TO CONTINUE
                  IF ANSWER = # THEN
                          CURRENTPAGE = CURRENTPAGE + 1
                          GOTO DISPLAY .

                  IF ANSWER (STARTS WITH) # THEN
                          CURRENTPAGE = ANSWER (WITHOUT #)
                          GOTO DISPLAY
                  *    RESPOND FOR 1 LEVEL DEEPER
                  CURRENTPAGE = PREVIOUSPAGE (ADJACENT)
                  ANSWER
                  GOTO DISPLAY .

NOPAGE            *    PAGE NOT FOUND IN LIBRARY
                  *    TO AVOID DOUBLE CHARGING:
                       RESHOW = 1 .
                  MOVE MESSAGE 'NOT AVAILABLE' TO
                  MESSAGEAREA .
                  GOTO DISPLAY .
```

Table 14-1 (continued) Base Code for a Viewdata-like System

```
BACK                    *    BACKPAGING CAPABILITY
                        *    BACK FROM INITIAL PAGE = EXIT SYSTEM
                        IF PREVIOUSPAGE = X GOTO EXIT .
                        CURRENTPAGE = PREVIOUSPAGE .
                        *    TRUNCATE PREVIOUSPAGE TO NEXT HIGHER
                             LEVEL
                        PREVIOUSPAGE = (TRUNCATE 1) PREVIOUSPAGE.
                        GOTO DISPLAY .

EXIT                    *    TRANSPORT ACCOUNTING DATA TO
                             ACCOUNTING SYSTEM
                        USEDTIME = SYSTEMTIME - STARTTIME .
                        WRITE USERID, USEDTIME, PAGECOUNT,
                        FEE TO ACCOUNTFILE .
                        DISPLAY EXITPAGE .
```

---

rewrite it in seven colors and graphics. The display editing system (DES/CMS) was used to create and maintain pages.

The example demonstrates how easy the writing of a basic videotex system can be. Even so, unless one is a computer manufacturer or software firm, better buy than make. Then what is the reason for including this discussion in the introduction to this chapter? The reason is insight! Although the system to be employed for management had better be developed by videotex professionals, the specialists of the user organization stand to gain by applying themselves to writing some simple routines. When they understand how the system works, they will be much more efficient in providing the interfaces. In that sense, in the following discussion we will review some of the reactions people have had to a hands-on experience in this area.

Writing Videotex Software

Videotex software written for commercial use is more complex than the simple example we have considered. A discussion with the designer of one of the packages revealed a significant insight on the way to project the interface to the database kept on mainframe—while respecting the VDB functions.

Database construction and update routines took one man 3 months to write (while not working too hard, by his own admission) and incorporated:

1. User passwords and privileges
2. A utility which prepared an empty DB, provided a dummy directory

(empty pages), cleaned and prepared empty page records, and, assured page management capability (to modify, add, and delete)

The retrieval program takes a 24K memory, and retrieval takes less than 1 s; but it takes 4 s to fill the page. To accelerate man-machine interactivity, the subject routine was written to transmit data in three layers, with checkpoints for "next command" interrupting current processing and moving to the following page being called by the terminal.

The advice on how to proceed with similar problems is based on practical experience by the software specialist who wrote these programs. It should, therefore, be looked at very carefully. Even if a point is of no interest in a specific application, the lessons learned through direct experience can be most valuable. Those lessons start first and foremost with the procedures to be followed in formatting the VDB. They include:

1. The need for establishing standards on page design and page access

2. The steps necessary in creating pages: from scratch, changing lines (for example, altering a financial value), and inserting color

3. The enhancement of terminal compatibility: retrieving pages and modifying them

4. The requirement for standards to be established for coding sheets which will be needed for original design, page maintenance, and getting new users into the system

Quite evidently, the greater challenge lies in extracting from the master files for videotex presentation and the bulk editing requirements associated with that operation. Here the key seems to be in the careful establishment of report formats to be handled through a parametric routine. The advice the software specialist with the videotex experience gives us is to write a small supporting program with a limited vocabulary to edit report by report. Whenever there is a change of format, a *template solution* can be of help.

A certain amount of ingenuity has to be shown when converting hardcopy report formats, classically used by the company, from the 132 columns of printed form to the 40 columns of a videoform. For instance, long clear text can be put in the middle of the line as a subtitle, and color can be used to differentiate normal rows from subtitles. Still, this may not be sufficient for reducing a form to 40 columns. Then the next step is to establish the fields we are interested in, and provide a set of values which uniquely define the field of interest in that report.

- Is it a header line?
- Is it a subheading line?

- Is it a detailed line?
- Is it a totals line?

These are the basic categories in a report. "Report preparation for Viewdata," the specialist said, "looks very much like RPG." (RPG stands for report program generator and is a programming language.)

In the case of columns, a subheading reflects the structure of fields immediately lower, but the structure of the fields is not necessarily the same as that of the subheading. Taken together, rows and columns will create the structure of the *output* form in which we are interested. (The term "output form," or "file," is synonymous with the format of the pages which constitute the VDB.)

For the design of the output file we should first establish a set of definitions which correspond to input conditions. Having an input and an output structure, we may wish to support more services. That calls for a *fixed procedure* able to interpret I/O specs and be independent of the formatting of the page but be controlled through parametric approaches. (For the input file, this approach takes a batch spool. The methodology is, however, valid whether we talk of spooled data or direct access to the updated master files.)

Now let us recapitulate: To proceed with the input in an able manner, we need input structural definitions. Hence, after categorizing the structural elements of a point file into the four classes, we have to qualify them. This means:

- Position (column number, field size, attribute, numeric and alphanumeric)
- Qualifier (if columns 1 to 5 are blank, then the field may exceed preset size—for totals)
- Exception identification (through color)

Values in the input file may be subject to moving across *as is,* dropping, or manipulating, as by creating summaries and providing datascreen footing. Answers should be given to these prerequisites through the I/O specifications. Some rules can be helpful: For instance, when a field moves from an input to an output report, give them the *same name.*

If we succeed in establishing some fixed logic in the overall procedure, that will be quite a positive result because rules will apply:

- Through the input report
- In reading record by record
- In identifying the type of data online
- In moving text and data to the output form
- In positioning on datascreen formats

A fixed-logic approach will typically read records one by one and move text and data across by consulting the two templates (input and output):

1. Read input page.

2. Apply prepass if given code is present.

3. Move field (through template consultation).

4. Apply postpass (if the code is there).

5. Write output page.

Such a formatted approach can be instrumental in dropping text and data or can help create extraordinary pages to feed back into the system. It can also generate cross-indexing and directory (Fig. 14-1) and lead from input files to output formats. The fixed logic will move text and data across depending on structure and definitions. These definitions are expressed through the templates. They are specific to an input file, and they are in need of proper cataloging. Names should be used as defined, and we should avail ourselves of the capability to skip fields in input specifications (skips, dummies), insert fields in output specifications, and assure that the procedure has an exit point.

An indirect addressing approach was suggested to help implement read and write permissions. This is most important, because everybody working interactively has to have permission keys. Practically, such a facility can be associated with a log-on capability.

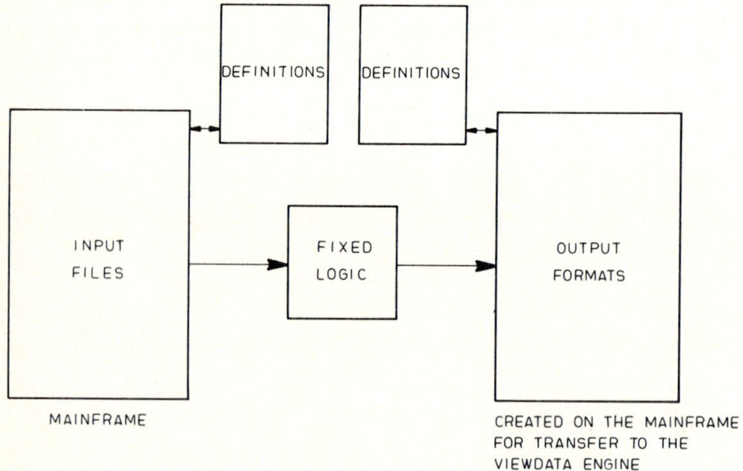

Figure 14-1 Using cross-indexing in leading from input files to output formats.

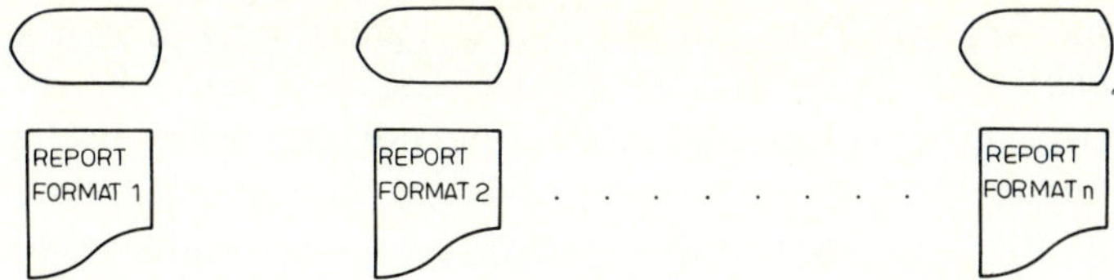

Figure 14-2 Carefully established softcopy report formats to
be handled by a parametric routine.

One of the characteristics which makes the organization of the VDB
quite different from that of the mainframe database is an *uneven* update
requirement: Some pages will be changed daily, others weekly, still
others monthly depending on the end use. To run an uneven update
algorithm properly, it is advisable to:

- Label the frames with the expiration period (sunset clause at page
  level)

- Make the frames open to read only up to a given date and then read or
  write

- Guarantee a period of nondestruction and also a period of update
  capability

- Include an established obsolescence clause

In a different sense, not only must a variety of video-oriented (softcopy) report formats be clearly established to be handled by a parametric routine, as suggested in Fig. 14-2, but also those formats (and their
content) should be subjected to sunset clauses. Videotex has the key
advantage that it has been projected as a lean reporting system. It
should not be allowed to collect fat.

Full and Partial Publication

To make the transition from their large, general-purpose databases to
the VDB as easy as possible, several users have been looking for rules
and guidelines. Among those I have collected are some that concern full
and partial publication. Although the parameters of an interface routine should assure the formatting of pages and enable the user to specify
the routing structure while the package generates the indexes, cross-
references, and similar facilities, we should not forget that, with a
management information system, only some of the infopages will be
changing every night. Others are of a weekly, monthly, quarterly, or
even yearly nature.

Updating takes time and has inherent in it the possibility of errors. Therefore, a mechanism must be available to assure that only the selected part of the VDB at the user site is regenerated every night. There are two alternative regeneration possibilities. The user can overlay or reload but must be able to give to the package a parametric instruction:

1. The *full-publication system* will renew the full VDB every time.

2. The *partial-publication system* can vary action day by day and replace frames that have been affected (Fig. 14-3).

However, if we change the routing tree in the VDB structure, then full publication is necessary. On the other hand, periodic full publication (for example, once per month) allows us to observe sunset clauses.

Partial publication presupposes the capability to reinsert records. The package keeps track of page numbers and also copies of the master file in a way similar to the INPUT (mainframe) form (the internal package version) which permits it to reassemble as necessary. This is a needed facility because if one screen is composed of, say, 20 records and only one changes, the package must have access to the other 19 to reassemble the new screen. The facility is also instrumental in assuring the proper physical and logical parentage. Logical parentage is the way of getting to the frame: free choice routing and cross-indexing. (Free choice means we can go from any frame to up to 10 choices supported by the routing frame. Physical parentage is the tree structure of the VDB.)

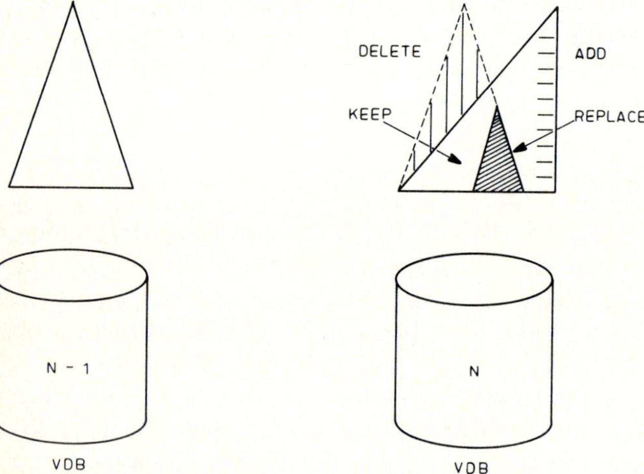

Figure 14-3 Add, delete, replace, and keep through a partial publication capability.

Logical key choosing can be used to advantage in going from a known starting point to a user-oriented organization such as an industrial classification scheme. Associated with numeric references (flight numbers; day, month, and year; part numbers), this can become a powerful mapping structure in parallel with the menu selection capabilities. A different way to make that statement is that, as skill in VDB design grows, users are demanding professional access routes without looking at menus. Some of the suggestions advanced in preceding chapters had that requirement in the background.

Another crucial issue concerning the updating of the VDB is the base from which the tabular data for the infopages should be extracted. It is evidently advantageous to use the master files to construct graphs through computer support, but that is not quite true of tables. Some users choose, here too, the master file reference; others let the batch programs on the mainframe prepare the output on spool and proceed from that point with extracting for and formatting the infopages.

But approaches have their merits—and their limitations. With master file access there is a limit on the kind of files the interface can handle. When sequential or index-sequential (IS) organization characterizes the files, the customer should use a report generator to produce a simple sequential file. The file will in any case be processed sequentially, and the user must describe to the package what the file will look like, select the record type, and identify the field size. I would tend to regard these specifications not as outright limitations but as good ideas to follow in the design of your own routines. The polyvalent interface which can do everything at the same time does not really exist, and if one sets out to create it, both the time delays and the cost will be prohibitive. Nevertheless, most packages support some flexibility. The file may contain fixed and variable records, and so may the fields. The description of fields can fit internal purposes, and the names are transparent to the end user.

Having planned the editing operations on the input data, it is advisable to define the rules to be applied to incoming fields prior to elaborating further detail. Formalisms are necessary to assure that editing will be only by programs and never by keyboard.

Personal computers offer new perspectives in these respects. Users have already started to employ microcomputers for accessing the corporate mainframe in a timeshared mode. They also acquire expertise on how to *download* data to their PCs.

Such activities can take place in a videotex mode or through data streams which are not formatted through paging. Videotex has a decisive advantage for the user who is not a computer professional. It assists in the correct interpretation and framing of text and data.

In this sense, videotex helps the PC user because it solves compatability problems in a manner transparent to him. Users can exchange documents and graphic images with equal ease. The necessary interfaces are run by the machine.

Most interface packages are rule-driven. As such, they inevitably have limits to what they can do. Beyond those limits we must provide programs which can be integrated to, for instance, sum up database subsets to provide summary data. However, a mainframe interface should be able to read in subordinate files taken out of the master file(s). To enrich information, it can use codes to link to other files. To edit (upper and lowercase conversion, for example) it must follow preestablished rules. Computers usually have capital letter printing; the package must make the conversion if the user chooses.

Layout Rules

The frame is made up of two parts: constant and variable text and data. When a screen is being set up, the general makeup should have customer approval before being described in the interface, and standards should be used so that all analysts can be in conformance.

Any end page is an infopage. (On Prestel, every page, with the exception of a response frame, is the same, but that is not true of Philips.) Pages are built into two parts:

1. Standard (pro forma, or skeleton) pages, which follow fixed contours of the frame that will not change

2. The filing of text and data onto the screen

The package should arrange transition between incoming and outgoing record types, given the different frame requirements. Procedures and software drivers are needed to provide control at the frame level:

- Records to be put on the frame

- Multiple-frame presentation

- Avoidance of breaking records between frames

- Positioning of text and data in fixed locations within a screen

- Provision of textual headings

Conditional-color rules must be supplied in the input (prior to the package) through a user exit; they are not supported by some interfaces and are native to others.

In formatting, the package usually builds up each record by using the start byte of a field. This is a helpful practice when text is variable.

Also, the package will use a "line-breaking" routine to segment. To do so, it has three components:

1. A report writer exploiting the videotex medium (to work on the infopage)

2. An automatic indexing system, which allows the building up of tree-structured access routes (for the index pages)

3. Mapping and output routines (page numbering and drivers for different videotex systems)

Indexing is the most important part of videotex, but it can be assured by using multiple key terms. The hierarchical structure will guarantee the "logical parents." The parallel organization, through two parallel runs of the interface, will also be hierarchical but will obey different criteria than the first (Fig. 14-4).

A more flexible structure can be obtained by specifying different record types and formats and also by cross-indexing. The routing page organization can represent the alternative choices. If a page has more

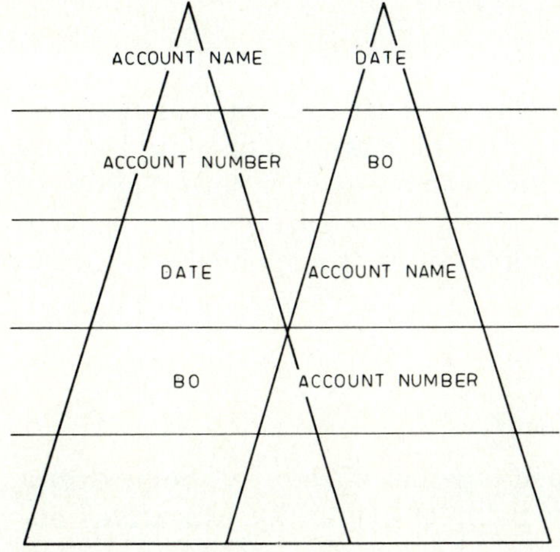

Figure 14-4 Different logical parents resulting in parallel organizations which are hierarchical in nature and composed of similar elements. The structures shown are the result of two runs of the interface package.

than two records in it, we can bracket them in the identifier (routing page) by referring to the first and last of the records. When more than one record is presented (for example, a corresponding bank's alphabetic index), we can use this bracketing technique to avoid splitting records or lines.

A well-organized system must maintain a choice. "Absent" frame numbers should not break the chain if "next to" is being asked. However, if direct access is made, the system may say: "Frame nonexistent." Furthermore, the interface package should be size-independent; it should be able to build a meaningful index no matter how large the number of choices. This facility can be supported by grouping into sublevels.

The stated alphabetic data organization is an example. An alphabetic organization can be in the sense of an increasingly refined search. An alternative to a sequence of finer alphabetic filters may be the use of two-digit numbers, a helpful bypass if we can contain the explosion to two-digit choices.

All these are tips on VDB structure. We can rephrase them. The first basic rule is to understand the application(s) in depth:

1. What is on the computer file?

2. What do we want to display?

To answer those questions effectively, we must look into the text and data first and then decide how we want to make the display. Display approaches are addressed to specific people and therefore may be in conflict, whence the wisdom of answering the second question prior to design. In short, we must first think through the application:

- Who will use it?

- What is the benefit?

- What text and data are necessary?

- How can text and data be grouped together effectively? Some examples are by industry group, by point of origin, by executives addressed, and geographically.

Answers to these queries make it feasible to protect a screen mock-up and develop a sample index frame. Analysts should simulate the expected results from the approach they have chosen: see how they work out and, if need be, set up a different approach prior to making structural decisions.

Anticipating access routes depends on the data being indexed. In that regard, some fields are easier to structure than others. For books, the access is by author, title, classification, editor, international standard book number (ISBN), and date. But classifying advertisements, for instance, can involve a far larger number of indexing possibilities, and projecting the range of end uses helps identify the best routes of info-page access. And it is surely advisable to consult the end user on the way.

Identification is another challenge, and the use of "numeric synonyms" (pseudo-mnemonics) does not seem to create any foreseeable problems. We must, however, be careful in allocating numbers so as not to run out of them. (A *unique* item should be an a frame in the Prestel jargon. If we need more frames, then we should use b, c, and so on, but should not push too far down the alphabet because of the difficulty of accessing these pages.)

Indexing may become uneconomical if we make long trees. We should be most careful to neither over- nor underindex: Too many layers of indexing pages in a routing tree lead to wasting the end user's time, and they constitute an example of underindexing. As an extreme, we can have only two choices per routing page. Then to create, say, 20 different options, we would need 5 consecutive layers of routing pages (2^5, which really gives us 32 choices).

Even 3 options per routing page underutilizes the machine's branching capabilities and should be avoided unless structural reasons oblige this solution. Typically, a routing page presents 0 to 9 possibilities. But it is poor practice to use at the beginning all 10 possible choices; even 9 or 8 of the 10 is too much. If we do overindex with, say, 10 keys on a page, we can, with 2 consecutive pages, cover the 20 options we took as our example (10^2 really gives 100 possibilities, but we said that we need only 20). However, the problem is that, if we needed another choice on the page, there would be no room for it and we would have to restructure the routing tree.

Restructuring the routing tree is bad practice because it disturbs the videotex user by altering the menu selection-based access route which he has learned. The best approach, therefore, is to start building the menu by employing 5 or 6 possibilities on the routing frame and exceptionally only 4 or 7 if the situation warrants it. With 5 or 6 choices we can still contain 20 options on two consecutive pages ($5^2 = 25$; $6^2 = 36$), though the maximum number of options is less than that we would get by using all 10 lines per menu page.

It is proper to emphasize at this point that criteria other than filling up the possible choices should be used in designing the menu frame:

- Homogeneity of the information contained in terms of options
- Logical grouping of choice issues

- Organizational prerequisites in sectionalization
- Established procedural paths we choose not to alter

Although menu selection is a nice way to break in the end user, eventually he will get tired of it. Users who learn how to work with the system soon get impatient with having to channel through successive routing pages all the time. They would like to have *direct access* to the infopages they need, and the videotex system software should give it to them. That suggests the need for a careful design of the menu selection mechanism. It comprises layouts that are to the routing pages what the "frame skeletons" are to the infopages: standards that the designer can define and work on to assure that there is no risk of confusing the end user.

Furthermore, we should carefully choose the parameters to be employed when the application becomes operational. (Some 30 to 50 parameters are typical; up to 100 are possible in complex applications.) We should then validate parameters: mandatory fields, numeric values, and so on. After validation is the time to compile a clean set of parameters. If the parameters are in error, the data will not look the way they are supposed to look. With expertise, constructing an interface package may take only 2 or 3 days starting from scratch. But the intellectual, preparatory work may demand several man-weeks.

Videotex calls for skill in looking at the total picture. The user organization should not take only a very limited view and focus on a small part of the VDB. The medium is new, and it demands new perspectives.

User Exits

There are some other rules to be applied in the design phase to assist in the projection of a valid system. Several relate to the user exits, including the handling of discrepancies and page breaks. User exits are required, for example, when the input file is in special format (or comes from an alien computer) so that a standard interface package cannot handle it and a special routine must be used. Another need for user exits is when the interface package can read input file but we wish to keep control of add, delete, and modify. Still another is related to the sorting facility on different manufacturers' software.

The ability to insert synonyms, establish a free text retrieval capability, and enter a stop list of nonmeaningful words to be deleted can also lead to user exit requirements. The free text retrieval capability is related to office automation. Correspondence files and memos may well call for user exits. The same thing is true of receiving control after page makeup, page numbering, form manipulation (mainly for replacement purposes), or special page numbering for alternative indexes.

User exits may well be necessary for the incorporation of graphs with optional media conversion or incorporation of cells. They may, for example, call into play a Fortran program to create histograms, bar charts, and trend lines (in single color or multicolor) and then proceed with the necessary sort and merge utilities to incorporate the graphs into the VDB.

Page-break rules can also be supported through a user exit. There are alternative ways to start a new page:

- Put a record limit on it.

- Start a new page (or frame) with every record.

- Start a new page when index terms change.

- Base choice on criteria the package offers; give the order through parameters.

- Switch to a new page when data conditions change.

A generalized mainframe interface does not necessarily handle all of the possibilities; in fact, it may handle none of them. It is then up to the user program working through the appropriate exits to the main routine to put the selected page-break rule into effect.

The subject of user exits is most important when the organization contemplates buying a commercial interface package. Just because, as already stated, we are at the beginning of this art, management can be sure that what is commercially available at this time will rarely fill all its needs. It is, therefore, most advisable to keep open options for further expansion through user exit capabilities.

Graphics, for example, may well be the first candidate for the use of the interface routine's user exits, but in a more general sense the interface to be chosen (or made) should assure:

- A direct connection to the batch database

- Full observation of the site's constraints

- Type of data organization (sequential, index sequential, etc.)

- Edit proper (parametric) and its requirements

- Needed capabilities: color, exceptions, hash, and so on

- Graphical presentation oriented to the choice of coding (alphamosaic, alphageometric)

- User exists

- Assurance of communication between the mainframe and the videotex system: (floppy, tape, cartridge, online)

15

The Updated Text and Database

"The price of power is responsibility." WINSTON CHURCHILL

Prestel, the public Viewdata utility launched in England, has served its original purpose: technical validation. The system functions, and it functions well. Furthermore, the licenses sold to the German, Dutch, Swiss, and Hong Kong PTTs, plus some business firms, indicate a good penetration in the international scene. But in England itself, investments run high, and it is quite clear that they cannot be recovered in a year or two.

Those who feel the pinch on the financial side are the information providers. They divide into two classes: major organizations and smaller, private firms. Barclays, the Stock Exchange, the *Financial Times*, British Rail, and so on gain so much experience by participating as IPs in Prestel that, whatever it may cost, the experience is a bargain. That is not necessarily true of the independent IPs—the public relations firms and service outfits that depend on the consumer's pound for their existence.

To put it bluntly, for any IP that does not have a comfortable R&D budget, the problem is this: No one is making money out of Prestel and may not make money for another couple of years. There are, on the one side, the known problems with start-up costs. What airline profits on a new route the very first year? British IPs also see some system problems. Prestel, at the time of writing, supported neither message facility, sophisticated reporting, text processing, calculation, nor data processing.

Videotex systems in other countries have learned from the BPO experience and have broadened their range of support. Still, I can easily understand the feelings of a British IP who said: "I am a businessman; involved with making money, not with burning up my financial resources. Most computer systems are complex and difficult. Prestel is simple. It should have got swiftly off the ground—but...."

Although the people who invest their money as information providers to the public utility are evidently looking for a return on investment and may get a little nervous if they do not see it coming, it is equally evident that the users of the services, who can have information at their fingertips through an easy, TV-type access, often free of charge, feel differently. To show the diversity of possible information, Fig. 15-1*a* presents as an example home and business services (this is a multipage), Fig. 15-1*b* is an index frame for time out/night out, and it is followed by infopages, Fig. 15-1*c* advises on rates commanded by night life viewframes, and Fig. 15-1*d* gives an example on museum information for New York City, which costs 5 pennies (about 10 cents).

There are many facets to consider when we talk of new types of services. First, even if the public utility service is not yet a financial

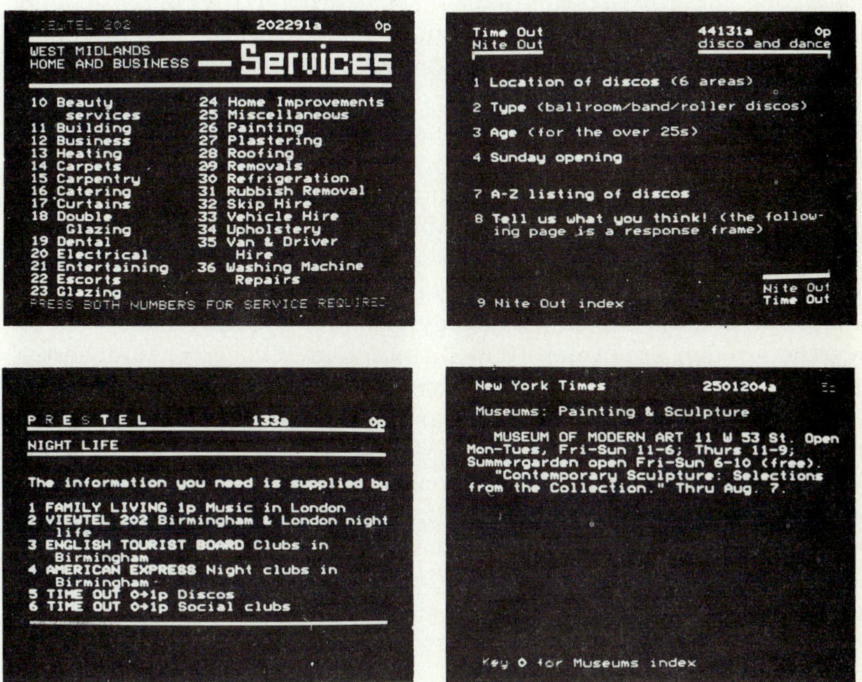

Figure 15-1 Four examples from Prestel-supported videotex pages.

success, there is a broad (and, most likely, lucrative) market for in-house systems. Second, within the public utility itself there are several interested sectors, some of which may prosper better than others.

That is the thesis of the IP who said: The BPO will get the public market. The private Viewdata system will get the major users. This may not leave much room for umbrella-type information providers who make their business preparing and providing Viewdata pages for other firms. but those industries who use Prestel as a laboratory experiment are sure to make a profit.

The In-House System at the Stock Exchange

Several British industries have joined Prestel as IPs, a proving ground of interactive experience. As data processing users move out of the middle ages of batch into the area of man-information communication, with word processors and intelligent copiers imposing another range of needs, the knowhow is still thin. We all know that there is no better way to develop human resources than by real-life experiments. Such experiments have been valuable in learning how to use a videotex system effectively, not only in terms of technical characteristics but also, if not particularly, in relation to the organizational aspects. It is, therefore, quite important to consider at least one of these experiments both on its own merits and in relation to the aging system for which it substituted.

The London Stock Exchange was one of the first IPs in the public Prestel offering, but it had already been using, for nearly 10 years, monochrome TV receivers to display stock market prices and news. This service was radically restructured along in-house videotex lines. The original system, market price display services (MPDS), served about 1800 display units located in the central London area by using broadcast cable television technology. Another 150 receivers were outside the London area (mainly in Birmingham, Glasgow, Manchester, and Liverpool). The out-of-London connections were made by transmitting through phone lines with a local PDP 11 concentrator and coaxial converter.

This system handled 800 quotations with market price indicators by sector (insurance, banks), mainly British and some Canadian. There was also options market handling modeled on the Chicago Exchange. The indexes used were those of the *Financial Times* and Dow Jones. The system also included three pages of company news such as interim results, profits, and sales.

The database was regularly updated, with quotations changing every minute. MPDS worked well, and the results it offered were commendable. But like all information systems (good or bad), it was getting short of space. More space was needed for:

- Overseas securities
- Traded options
- Selective presentation for brokers and their clients

It is, therefore, no surprise that the London Stock Exchange thought of alternatives. The use of teletext was examined as possibly transmitting faster than Ceefax. But, as is well known, this service lacks interactivity.

The teletext study did, however, give the exchange valuable information on the selection of terminals. The tests were made with a mass-produced, low-cost TV terminal, which could be distributed cost-effectively to the member firms. This helped establish these criteria:

- The cost per unit should be £850 ($1350) as a maximum, not the $2000 quoted by some suppliers.
- The video terminals should feature color and graphics.
- The terminals should support multiple keyboard and printer.
- The terminals should be microprocessor-implemented solutions.

The British suppliers did not seem to have matched the requirements. STC and PYE offered a business videotex terminal but only in black and white. The stock exchange application must have color. The reason for this insistence is realistic. On the board the traders mark in color.

- Blue for a price going up
- Black for a starting price
- Red for a price going down

And the stock exchange wants to give on video the same image. Furthermore, Prestel and other experiments have shown color conveys much more meaning.

The other basic decision had to do with the wisdom of an in-house system as contrasted with a public utility. Prestel was examined in the light of a number of factors:

1. The cost to the stock exchange and its members.
2. Operability. The stock exchange must be plugged in all day, which the BPO does not allow.

3. The updating characteristics. In its original release, Prestel did not update the page once it was on the screen.

The updating limitation brought up the need for re-requesting, which is inconvenient. That and other considerations finally made the London Stock Exchange decide to start from the beginning but keep the videotex principle. In June 1980 the exchange put on stream the new videotex with up to 2000 color display units supplied initially. Each unit has a dedicated high-speed channel into the central computer system, a Modcomp Classic 7870. The system has been engineered to provide immediate responses to information requests, typically within no more than 1 s under heavy traffic conditions.

Subsequently, a second computer was installed for program development and backup purposes. The system architecture allows for significant growth of the display terminal population by plugging in more computer resources, and it maintains compatibility with Prestel and Ceefax/Oracle (teletext services) to allow users access to information on these and other compatible systems.

One of the challenging questions was how to drive more thn 1000 terminals through a mini. The specialists decided on a microprocessor-controlled statistical multiplexer able to handle eight terminals feeding into another eight-port multiplexer. Still another challenge was the capability to distribute on an office center basis. For this the specialists use MUX by office block. (Within London a private 9.6-kbps line is supported. The exchange building has separate lines to four telephone switching centers, which is a good design for reliability. For the rest of England, private lines at 9.6 or 4.8 kbps will be used.)

A further challenge concerned bottlenecks. If, say, 16 terminals operate simultaneously and ask for the same page at the same moment, how will the system react? Tests demonstrated that the processor supports page requests every 5 s (any page), and this provides reasonable assurance that, given the applications stream, there will be no bottlenecks.

The system employs TV receivers, which are adapted to display text and mosaic style graphics in a standardized form. Characters, graphics symbols, and backgrounds can be displayed in any of seven colors.

- Information is encoded by using a 7-bit plus parity character code based on ASCII.

- Text and data are presented in 24 rows of 40 characters.

Other performance criteria were that the machine would be required to store at least 100,000 videotex pages and provide, on demand, any page to a population of at least 400 terminals. Each terminal would call for a dedicated channel into the system which could operate at speeds of up to 9.6 kbps with 95 percent of all cases assured a response time of less than or equal to 1 s. (Response time is measured from end of request to start of screenful.)

Serving the Broker as the Ultimate Consumer

The goal of any information system is to serve, and serve well, the end user. That is how the system pays its wages. Computers and communications are means to the end of better, accurate, and timely information; and if such projects failed in the past, it is largely because users did not care to acquire the necessary experience.

That is exactly what some of the original IPs tried to avoid with Prestel's Viewdata. The London Stock Exchange started with Prestel and used it as a development tool to gain experience. The Byron computer was the hub of a computer-to-computer link. Manual keyboards were used for data entry (the way the brokers work with the in-house system), and the VDB was updated three times per day with the price variations.

The best proof of how the user regarded the system was in mid-September 1979 when the BPO changed the computer hardware. The brokers on Prestel called the exchange to ask: "Why did you not update prices?" (A change in computer hardware usually causes a temporary suspension of service.) Furthermore, just because the ultimate user likes Prestel, the exchange started to get revenues.

Against that quite positive background, TOPIC (teletex output of price information by computer), the in-house videotex system of the London Stock Exchange was developed. A primary goal of the TOPIC system was a development program to keep costs to an absolute minimum, mindful that the system would have to compete with the low-cost broadcast technology employed by the existing market price display service (MPDS).

In a nutshell, the TOPIC structure consists of data terminating equipment of the ultimate user, whether local or remote, feeding through private lines into a submultiplexer (SMPX). In turn, 8 or more (up to 16) SMPXs feed into a multiplexer which is run by a specially selected central processing unit (CPU). For any practical purpose, the CPU is a fairly large, fast, front-end computer. It acts in this capacity to the exchange price information computer (EPIC), which is the host or, more precisely, the rear-end computer dedicated to run the VDB.

EPIC consists of two PDP 11/70, one online and one standby, and it is linked to Prestel. Two EPIC machines are hooked up to the video converters and services such as *Telekers* and *Extel*. The main functions of this system are switching (for data distribution) and database management, which should be the function of any and every mainframe in the 1980s. TOPIC relieves the processors of the communications-intensive activity which is at an ever-increasing pace.

The machine is expected to begin writing a page within 1 s of its being requested and to maintain that response time when all terminals connected to it are active and generating page requests at a rate of one every 5 s. The system is also required to accept updates to the page file at a rate of at least four per second and to rewrite any displayed page which has been the target of an update.

The stock exchange technical staff undertook extensive evaluations which could meet the requirements. Performance was verified by sophisticated tests using two computers back-to-back, one generating information requests and the other responding to them. The tests demonstrated that the projected solution responded to the requirement, showing that up to 1000 terminals could be handled by one machine with the projected response time. The adopted configuration was a fully redundant dual-processor system capable of very rapid recovery in the event of failure.

Product planning for the new in-house videotex service proceeded very carefully, as it should. The specialists made provisions for expansion. System design had to assure that more capability would be available to hook up to the in-house computer system provided all went well. One of the projected service extensions was to allow the handling of information on the broker's client and his portfolio, as well as the evaluation of the client's holdings.

Such operations can, of course, be served both online to a central resource and through local intelligence because it is available and inexpensive. A personal computer looks like a good possibility for application on a decentralized basis, but the designers at the stock exchange have been well aware all along that any distributed intelligence at the end user's level must be compatible with videotex. Add-on functions are evidently possible, and some of them have been projected since early planning. There is, for instance, the possibility of putting on the system information, not available in the London Stock Exchange, about other stock exchanges, horseracing results, and commodity markets.

The exchange also needs facilities for authorization, special access in dial-up, and other purposes such as billing. Changes, particularly in software, may be required if the distinction between brokers and jobbers which now exists in England one day disappears. The current system is single-capacity, but a double-capacity solution will be needed if jobbers and brokers can do both tasks. That will impact on TOPIC, which should provide quotations and automate trading.

Further capabilities will be necessary as the British Prestel converts from a byte-based to a bit-based system. Rethinking will necessarily have two goals. The first is technical compatibility; the second is serving the broker and the ultimate consumer in a cost-effective manner. The cost of terminals and of the central facilities evidently weigh on the bill the end user has to pay, and one has to be careful about total costs. System solutions should be competitive while assuring a better service.

Accounting for Future Capabilities

The reference to the London Stock Exchange's designing the present system with future capabilities in mind should be given the attention it deserves. Nowhere is this issue more vital than in matters concerning data communications and databases. Future direction should always be kept in mind when text and databases are projected. Technology points toward:

1. Standardization
2. Terminal enhancement
3. System expansion
4. Communicating databases
5. Increasing online usage
6. Expanding user application

Standards are needed for mass production, distribution, and use of a service such as interactivity, which within a short decade will come into general use.

If one asks whether standards are observed today, the answer must be no. There are technical differences between the systems being developed in various countries. But there is consensus that standards are needed, and in the end the differences will be smoothed out and an international standard will be created to permit people of one country to access information resident in the VDBs of other countries. This process will take some time, but it is important that a standard emerge. That is particularly true because the videotex success is volume-dependent, as is the success of both terminals and pages. Putting a single page on the system might be reasonably expensive, but thousands of users make the process much cheaper.

Standards are also most helpful in future enhancements. Voice, for instance, is expected to be one of the coming developments of videotex, and so is linkage to portable radio transmitters. Another add-on is going beyond still frames toward moving pictures, and the same advance is possible with terminal-to-terminal communications. Recall that moving pictures are feasible today with teletext, and the interactive and broadcasting processes can complement one another.

Eventually, videotex might serve as a window into the mainframe (looking into Big Brother). And videotex display units can take on extra features for new fields of use such as:

- Education (self-teaching)
- Mail order and holiday brochures through videotex
- Storage and playback (combining with videodisks and personal computers)

All this adds up to a social revolution based on the information industries, with the real takeoff located in the 1985 to 1987 time frame. Evolution can be delayed through legislation—or the lack of it, and it can be advanced through the development of large, publicly accessible and reasonably up-to-date text and databases.

The existence of professionally organized databases is sure to impact interactivity, communications, the choice of applications, and the definition and characterization of both the IPs and the most likely users. Taking a public videotex network as an example, database producers tend to be organizations traditionally involved with the provision of information: merchandising firms, industries, banks, publishers, associations, universities. Their databases can be categorized as source or reference. Source databases are *factual* (or primary). And they must be complete and detailed. Reference databases are of a bibliographic nature; they have to do with physics, electronics, computers, and so on. They provide support to R&D, and they help their users keep abreast of developments in particular areas. Source databases are designed as aids in solving immediate problems, as aids to decision making, and as means of assuring support for planning and evaluation, operational activities, marketing, and so on. Both types of databasing are part of the online service industry.

As a point of reference, over 300 databases are now publicly available in the United States from nearly 200 producers. In addition, there are many thousands of fairly well organized databases belonging to private companies and government departments. Over 50 percent of the public databases can be categorized as business, economic, or legal; 25 percent are scientific and technical. Of the database producers in the United States an estimated 50 percent are commercial, entrepreneurial privately owned firms, from Lockheed down to $1 million companies. About 25 percent are nonprofit organizations.

In Europe, the users are mainly government organizations, universities, and large industrial concerns, and the emphasis is on scientific, medical, and chemical databases. There is, however, a trend toward more commercial databases. More of them will become available, and they will be easier to access. A major step in this direction is the inauguration of Euronet, which makes it possible for users to access databases in many countries.

Databases in the United States supplied half of the 300,000 requests for economic, scientific, and technical information from the EEC in 1977, and the percent share has not changed markedly since then. The EEC market is on the threshold of a vast expansion. By 1985, the number of information requests is expected to exceed 2 million a year. Euronet will give subscribers access to 180 databases run by 23 host services and known as DIANE (direct information access network for Europe). Euronet's expectation of 1000 initial subscribers appears to have been a gross underestimate, and capacity had to be hastily expanded to 2500. The potential market is now thought to be about 30,000, of which companies account for 70 percent. Each item of information requested costs an average of $20 for a 15-min call. Charges within the Euronet system are the same regardless of distance.

The evident truth behind database producers is that power lies in the hands of the information owners. Typically, database producers tend to sign distribution arrangements with online service organizations, who will increasingly compete for database distribution rights and for the user marketplace. Their pricing policies usually involve a combination of *up-front price, use royalty,* and *online service.*

Getting Ready

Clearly, the existence of well-planned, rich, and properly maintained databases, although a necessary condition, is not a sufficient condition for interactive capabilities. More has to be done by way of preparation, as we often had occasion to observe. Any serious enterprise has found that on its own, and the brief case study included in this section buttresses the evidence.

Say, then, that a leading contractor wants to establish videotex for its management reporting, which classically has been computer-processed and presented through tons of hardcopy. To make menu selection feasible, a tree structure is designed to meet management's information requirements. Three principal accesses are chosen:

- Reports
- Statistics
- Information levels

These constitute the principal references to the reporting structure under the heading *contracted work* (Fig. 15-2). Two routing pages are constructed. The first has three choices: financial reports, statistics, information levels (company jargon). The second has six choices, as Fig. 15-2 shows.

Suppose this company operates on the west coast: California, Arizona, Nevada, Oregon, Washington, and Idaho. Management wants current year results by state, and it also wants area summaries. In California, there are four areas of operation: San Francisco, Los Angeles, San Diego, and Eureka, plus HQ (Fig. 15-3). This outlines the routing pages of the third and fourth levels, following the *financial reports* line. At the end of the fourth level, the user has access to the infopages.

If we follow the index *statistics* (Fig. 15-4), we again reach greater detail down to the construction site level, but we may require more routing pages. For instance, for construction work regarding a nuclear plant a choice must be made between maintenance and new construction and civil and mechanical engineering. Eventually, the infopages (which are at the end of the tree) can be reached by taking either route:

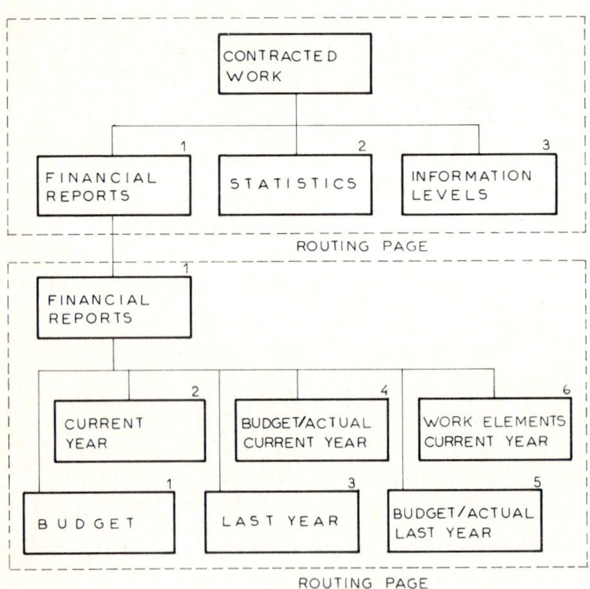

Figure 15-2 Two layers of routing pages leading to the financial infopages requested by a leading construction firm.

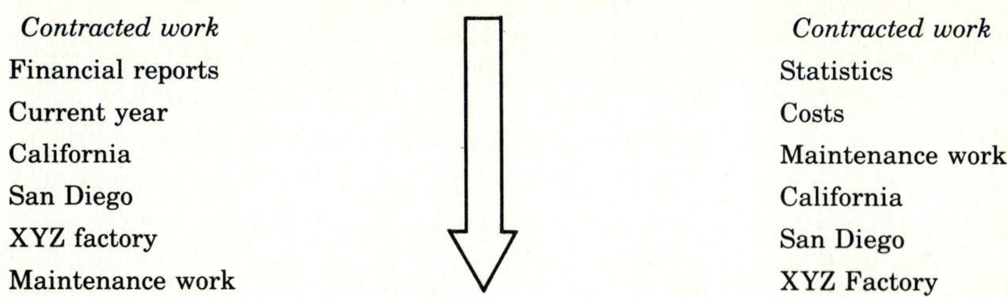

If the infopage at the end of the tree is the same in each case, it should be subject to cross-indexing. (Let us repeat what we have already said: the tree number is not a page number. It is a taxonomical consideration. The page should have a running number.) With videotex capabilities at his disposal, the management user can call up a page directly by its number if he knows the number. Alternatively, keywords can be used.

One more thought is in order here. In the background of this case study is an in-house system, but the situation would be no different, in

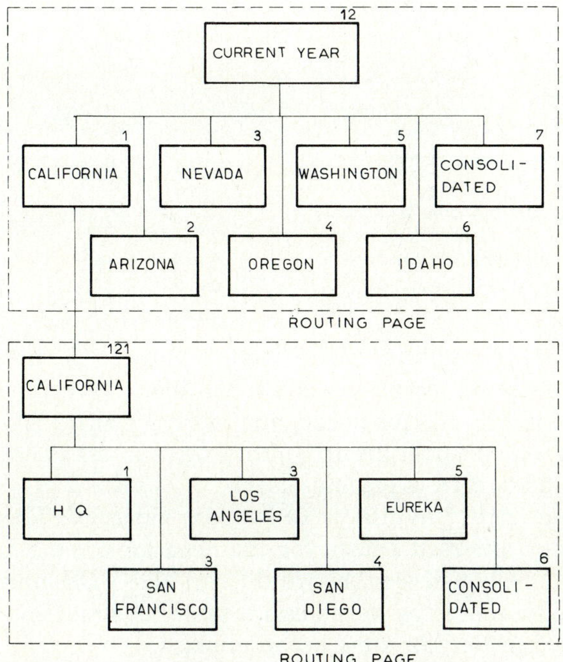

Figure 15-3 Based on routing pages, menu selection leads the user toward infopages designed to provide data on the geographical distribution of company operations.

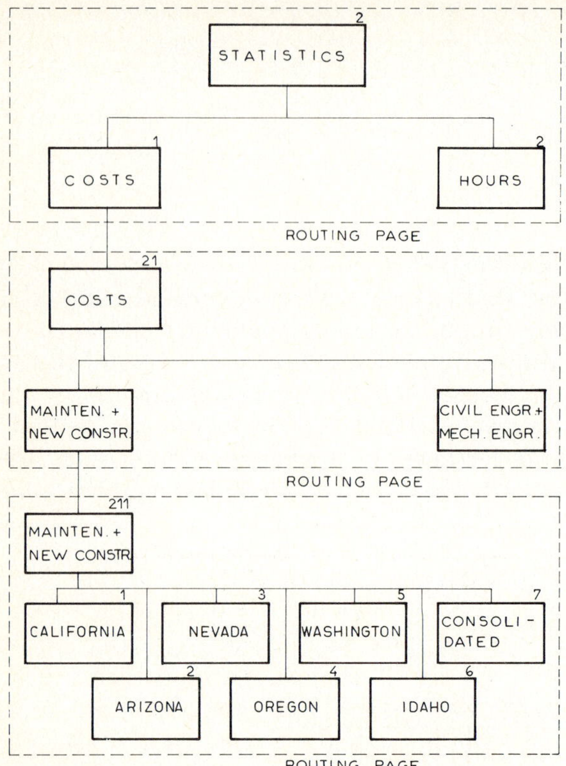

Figure 15-4 Statistics on costs made available by video-tex are organized according to the long-standing management method of presentation and hence cause under-indexing.

regard to user preparation, if a public utility were involved. In that case, the user would act as an IP, structure his trees (menu selection routes) in the way best suited to his purposes, and, with his state HQ and local sites, constitute a closed user group. In that way, access to his company's frames would be denied everyone else.

And if there is no public utility available to support videotex? The situation would not be very different except for the need for some FX lines. A centrally located in-house system would run the VDB and terminals at all subsidiaries, and sites would call it up. The versatility is impressive and the application horizons broad enough.

Access to the Central Information File

As another example, let us consider a financial institution in which the implementation of interactive capabilities through videotex brought up the need to review some aspects of batch processing on the host computer. One of them was posting directly to the general ledger, thus eliminating the need for the branch office daily statement to be compiled manually. Another significant timesaver was the automatic generation of list tapes for all the items that had not been truncated.

The availability of both the host computer information and the branch office mini with its videotex-structured format enables the bank to realize significant productivity gains. To analyze the learning curve of tellers being trained on the system over a normal working month, the bank simulated the five basic transactions which comprise slightly over one-third of the total customer activity. This major gain in productivity, coupled with the reduced balancing time, allowed the bank to reduce its teller staff and still provide a good level of service. As in most economic analyses of a terminal system, the personnel aspect is a most important factor.

Furthermore, in regard to the administrative terminals, the bank has made available to the branches all the information contained in its central information file (CIF). This data includes address changes, closing accounts, and other changes in the customer database. When a new customer comes into a branch, the branch establishes that customer on the CIF. As customers add new accounts, they are added to the applications files, and that information is automatically passed to the CIF.

With specific reference to the service to be given to its clientele (particularly the corporate accounts), videotex allowed the bank to institute a cash management system which combined management control with flexibility. Videotex income provides both realtime information reporting (last night's updating) and transaction control modules. Its functional elements are designed to serve as building blocks in the total integration of the cash management services which the bank now offers, very competitively, in its marketplace. This vital function consists of:

- Current-day reporter

- Cash reporter

- Regional balance reporter

- International balance reporter

- Money transfer input

- Target balance manager

- Depository transfer check reporter

- Foreign exchange reporter

All are organized into neat, well-defined pages as easy to call from the VDB as ringing up a phone number.

The emphasis placed on the administrative functions and the resulting benefits made the videotex application the catalyst in finding good solutions. The functions supported are as follows:

1. General accounting
2. Operations with corresponding financial institutes
3. General management accounting
4. Forecasts on liquidity and need for cash
5. Management of interest rates
6. Warranties
7. Credits
8. Portfolio
9. Commercial paper operations
10. Balances on foreign money
11. Client protests and solution steps undertaken
12. Regrouping of clients by category (industry, commerce, consumers) and type of bank relation

Furthermore, videotex pages have been dedicated to the task of managing the branch office properly, including personnel problems. Other services integrated with this system are:

- Wholesale and retail lockbox services
- Cash concentration services
- Disbursement activities
- Short-term investment services and long-term international treasury management
- Cash management
- Treasury management

The cash management system allows for a timely and accurate flow of information between the cash manager and the network of bank accounts. By monitoring the flow of cash across the network in a systematic way, the cash manager is positioned to use idle cash and maximize the flow of funds for investment purposes.

An interesting approach to retail banking and small business has been taken. The VDB provides each of those customers with a cutoff statement that shows all activity since the last statement. This service is of great benefit in resolving customer problems.

Finally, at the senior management reporting level videotex has permitted the addition of a broadcast function. This allows the central staff to transmit messages to all branches simultaneously. The messages are of many kinds. They update exchange rates, give instructions, answer queries, and warn branches about stolen or counterfeit checks, thus helping to reduce the losses from crimes of those types.

To achieve all this required decisions by management and commitment by the specialists. But because the videotex software is simple and standard, the development staff for a fully functional system was impressively smaller than what would have been required to design a private network and produce all minicomputer programs from scratch. And not only were the sizes of the implementation and support staffs required in the central and branch offices contained, but the operation was completed within a very short time. The branches were converted to the terminal system; the branch personnel were trained; and the problems were solved so that the work is processed correctly. "Videotex," management said, "has been a bold and imaginative step."

A Real-Life Example

In the videotex application described in the preceding section, some early experiences were integrated, but part at least of the success was due to a controlled experiment for the purpose of providing hands-on experience. The financial institute's structure for management reports is shown in Fig. 15-5. The main index has three-way exploration of accounting, statistics, and way of presentation.

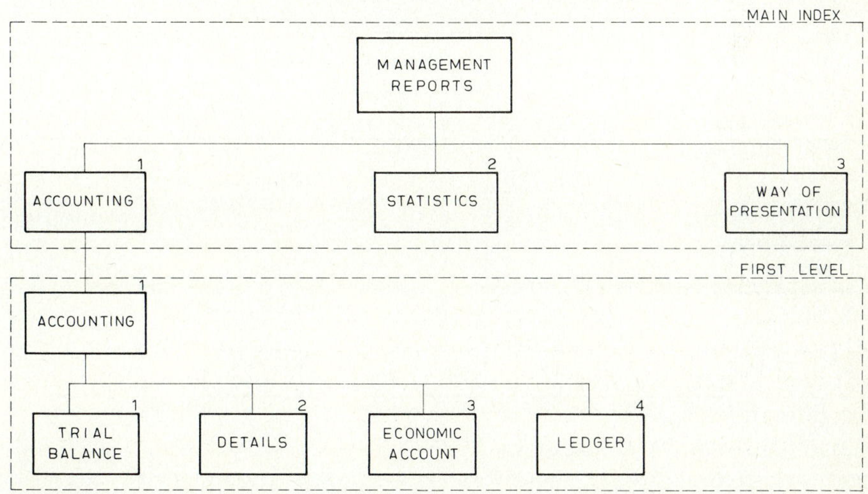

Figure 15-5 A management reports menu permits routing to the key accounting reports.

The first level of detail (the routing frame next in the sequence) involves a four-way split. It is followed at the second level (Fig. 15-6) by a list of 19 options. It can be handled either by making transparent to the user the required two levels (second and third) in the way we have described or by keeping it apparent through the regrouping of the 19 into a 4 by 5 two-level arrangement. (Other examples of two-key choosing can be month of the year, day of the month, or letters of the alphabet—a kind of telephone directory choice.)

The frame pertaining to branch office 1 was given the routing sequence 11012. But that frame may also be called through its direct dial number (say, 353-00) or through a keyword (say, Tucson). The first page of the multipage can serve as an index to be followed by up to 99 subpages which, however, should relate to it.

Figure 15-7 shows the levels following the choice of the statistics route. It will be appreciated that, from the first level onward, the organization is the same as that we have just considered, at the second and third levels. This is followed by a fourth-level exploration and finally the infopages identified in Fig. 15-6.

For an example of cross-indexing, say that the third path of the main index, that of the way of presentation, holds, at the combined first and second levels, the 19 regional offices (Fig. 15-6). The next level splits the route between accounting and statistics; followed by the exploration

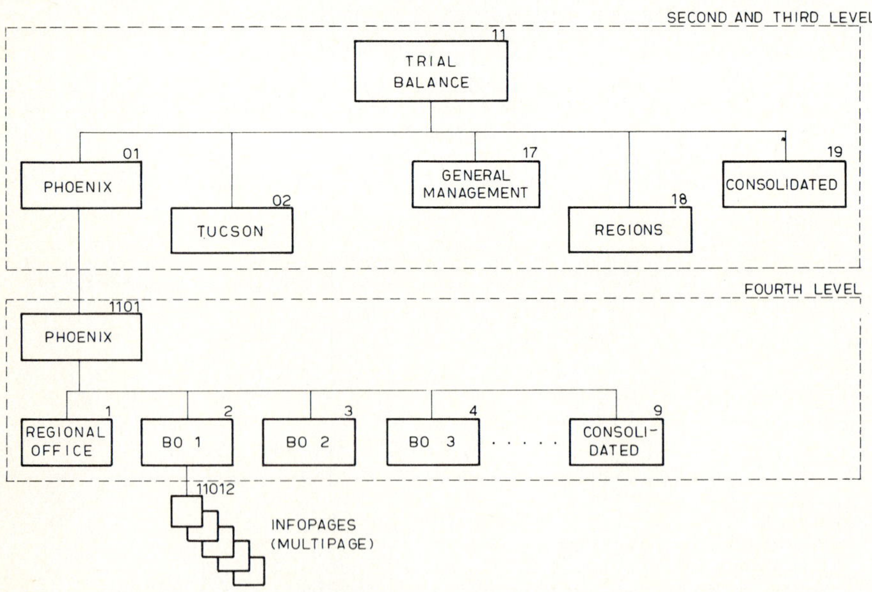

Figure 15-6 Second and third layers in the routing structure.

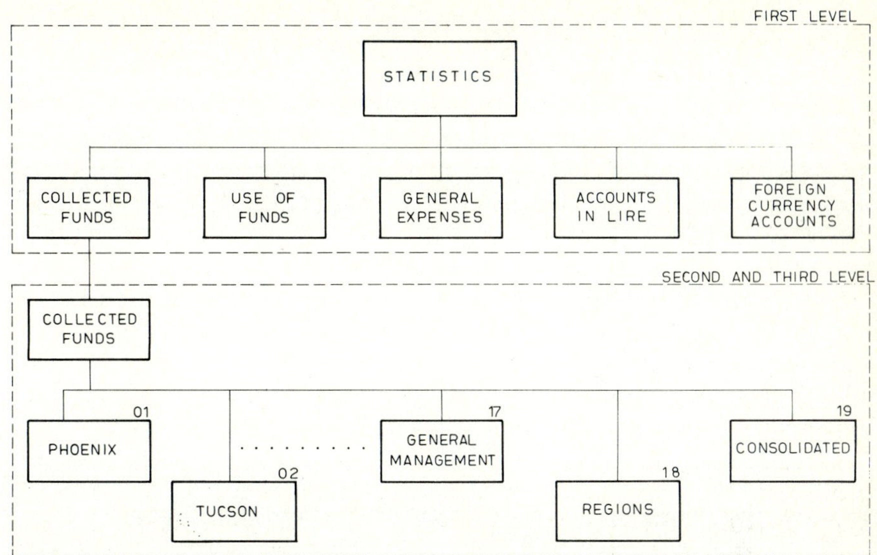

Figure 15-7 Statistics are presented through videotex in the procedural sequence followed by management with paper records. This has been a matter of basic choice.

into branch offices and infopages. With this routing, we have an excellent example of cross-indexing, as presented in Fig. 15-8.

Figure 15-9 shows the index tree (taxonomic organization). Five main references have been selected for management information, and

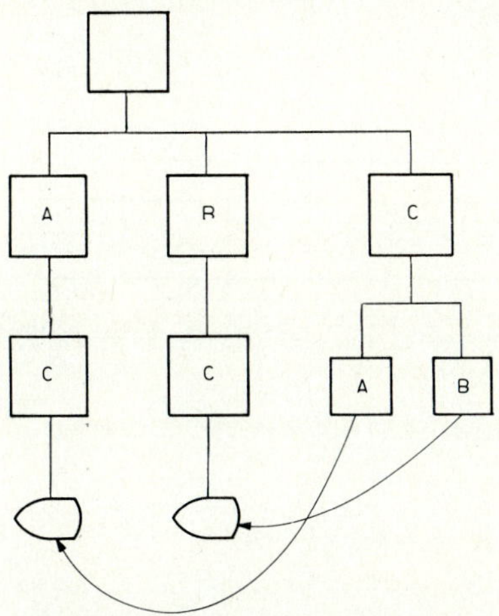

Figure 15-8 Cross-indexing can be easily achieved through videotex by linking infopages at the end of different routes in frame selection.

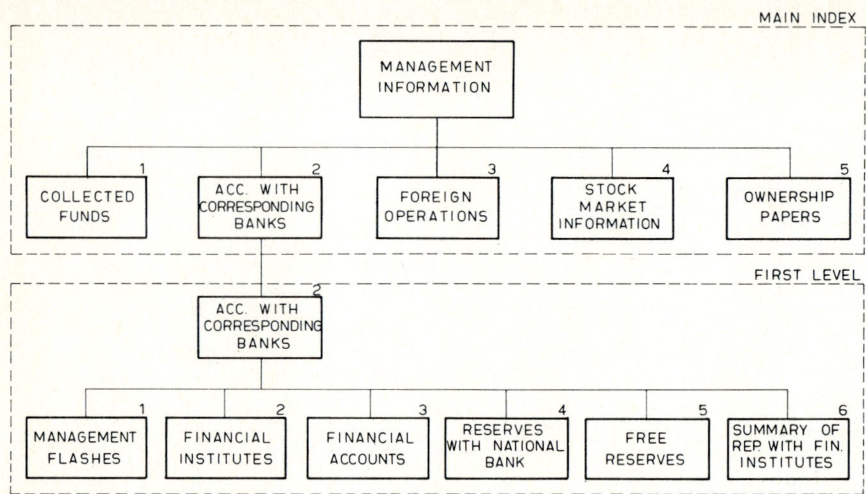

Figure 15-9 The routing for management information can be easily accomplished through a two-layered structure.

the example is followed with the option "accounts with corresponding banks" into Fig. 15-10. Finally, Fig. 15-11 presents a complete example with foreign operations distinguishing the levels of reference corresponding to the different routes and identifying the possible points of cross-indexing.

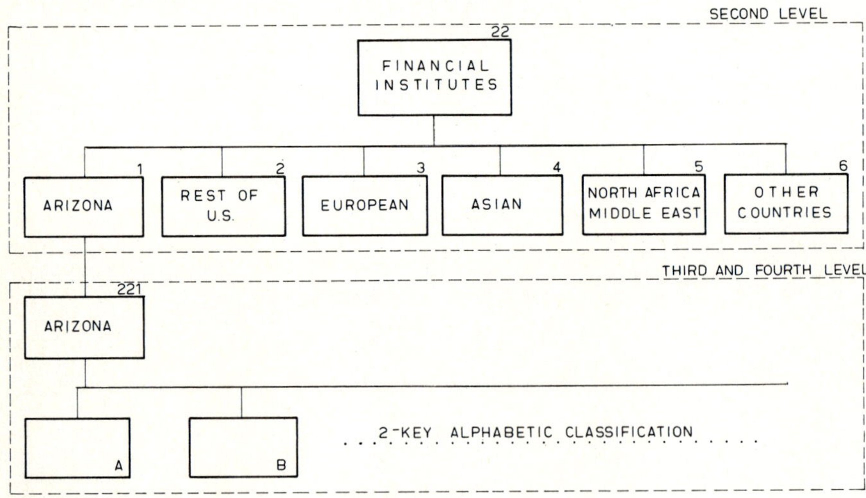

Figure 15-10 Menu selection possibilities may include two-key alphabetic search, provided the videotex software supports it.

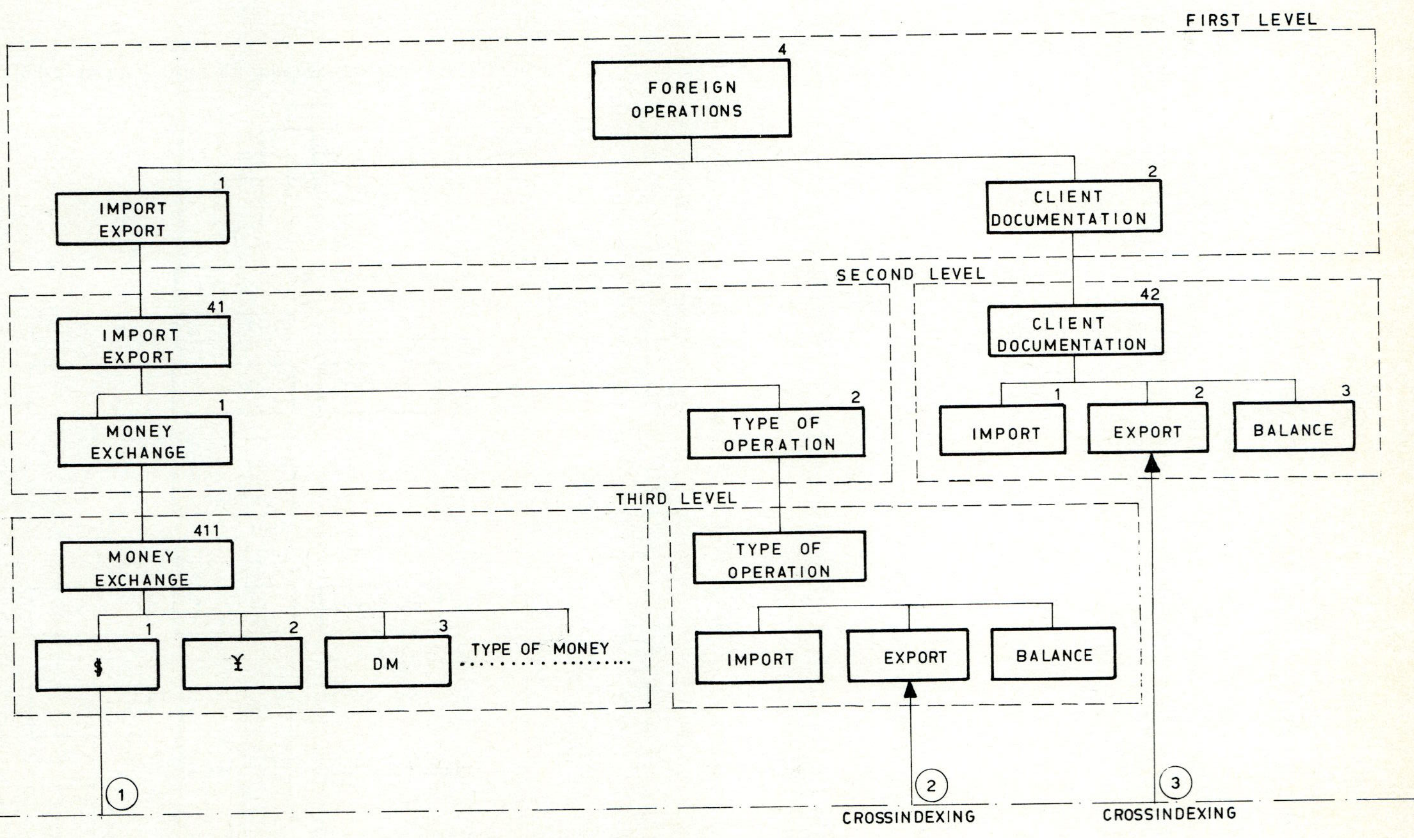

Figure 15-11 A foreign operations menu including type of money reporting, type of operation, and client documentation. The cross-indexing capability is important.

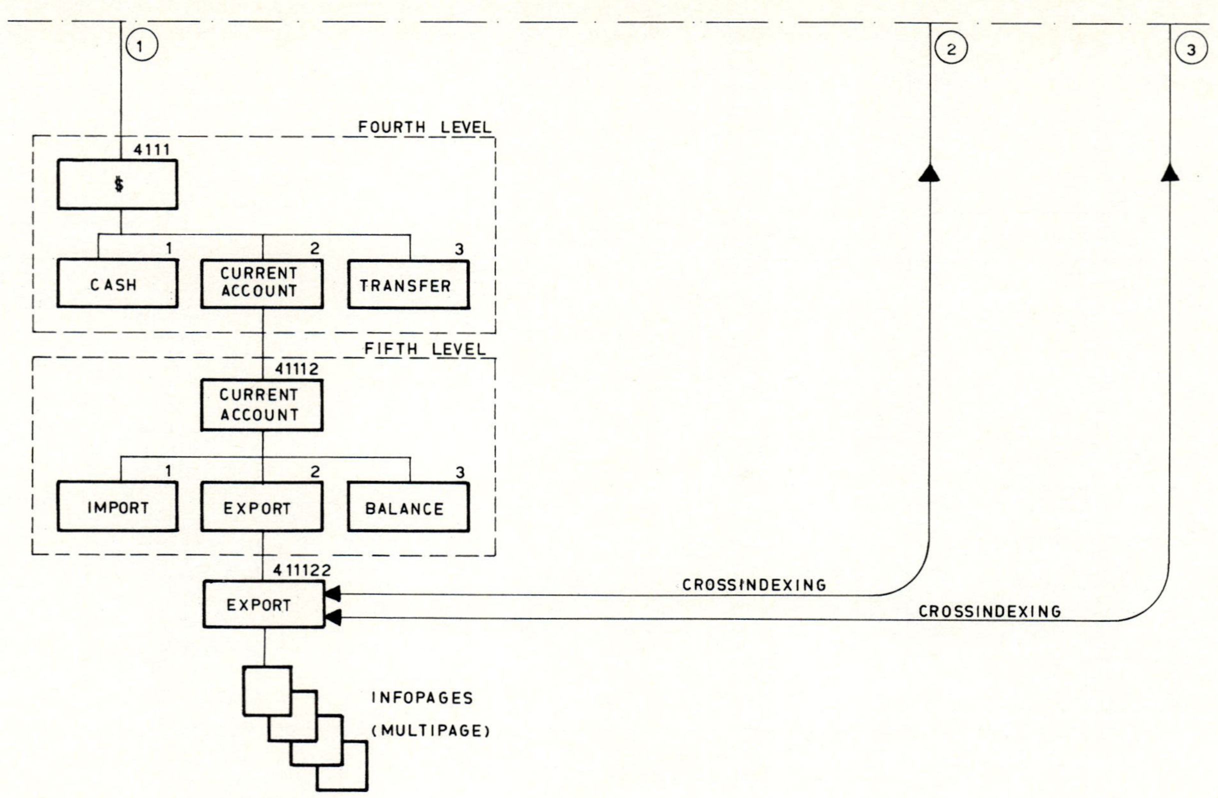

Figure 15-11 A foreign operations menu (*continued*).

It should be understood that this structure is still in evolution, but the results which have been presented help identify both the work which was done and the amount of preparation necessary to obtain a profitable outcome from videotex. Also, these examples demonstrate why so many frames have to be allocated to routing prior to reaching the infopages.

16

Home and Home-Office Banking Systems

"We promote them out here for efficiency—not age." Douglas McArthur

Chemical Bank, of New York, started developing Pronto in late 1979, and the project was completed in November 1981. In April 1982, 200 customers were randomly selected for a test, and by September that same year the service was launched. Statistics help indicate the amount of brainpower invested. In 1979, about 30 people worked on the project; in 1982, the number became 80, including the marketing people and the personnel of the software firm contributing to the project. The group may number more than 100 people in 1983.

Dedication is high. Everybody is young and enthusiastic; R&D, product management, quality control, and the sales group work hand in hand. The marketing people sell to retail customers in New York City and handle the licensing agreements with different financial institutions. As a result of this effort, Chemical Bank is able to market Pronto to correspondents. As with BankLink, management's intent is to license the software to any bank interested, even New York banks. As of May 1983, six American banks had signed the Pronto license.

In a parallel effort, Chase Manhattan Bank has four commercial banks and two credit unions providing 50 customers, at each institution, with the opportunity to test its Paymatic bill service. Southwest Bank, in Miami, has announced that it will provide gateway services to financial institutions who wish to participate in the Knight-Ridder Viewtron system. There is also the HomServ project currently being tested in San Diego. It includes such banks as Security Pacific, First Interstate, and Manufacturers Hanover. Other tests were made in Los

Angeles, Omaha, and New Orleans, and there is Bank-at-Home, a system offered by United American Bank.

A home computer, a telephone, and a TV screen, along with the software provided by the bank, is all that is needed to put Pronto and similar home banking services to work in customers' homes. And customers can avail themselves of banking services with an ease they have never known before, plus having a world of information and transaction services to come—all done quickly, easily, comfortably, and securely at home. Organized as a user-friendly menu selection service, this PC-based home banking service supports:

- Balance inquiry
- Electronic reports
- Funds transfer
- Bill paying
- Electronic checkbook
- Home budgeting

The Pronto service has been designed to give comprehensive instruction and guidance to the bank's customers. The customers can learn, step by step, how to use the available online facilities, and, as experience accumulates, the financial institution will be able to anticipate client reactions and prepare its staff to respond to them successfully.

The preparatory work should not be underestimated. To design an online system properly, intensive studies should be made of the capability of the bank's central information files to accommodate new and emerging services. Some procedures, such as the procedure for remittances to merchants, must be streamlined. Having customers enter payments through a telephone while the backroom people of the bank manually cut checks defeats the purpose of automating the system. The goal should be to provide greater customer convenience while increasing the efficiency of the operation.

The benefits can be significant. Among the fringe benefits provided the users of Pronto and similar offerings is an electronic mailbox. Subscribers can send messages to one another. The process is faster than mail and more efficient than answering services: Messages are stored for retrieval at the convenience of the recipient. As this service spreads among Pronto users, the bank can communicate with its customers without letters and calls. Management can broadcast marketing messages, address groups of customers through narrowcasting, and contact individual customers selectively.

For routine communication, the financial institution can save more time by using Pronto's standard message dictionary. It can also motivate shopping and other information services to market via Pronto. In that sense, electronic mail serves as an online link to the bank's customers. The latter can respond easily, communicate quickly, and handle situations with electronic efficiency.

Pronto provides the bank with complete audit trails, so that management can monitor transactions and locate any problems. Available software keeps the system up to date on its own operation by providing reports on transactions, changes, and exceptional situations to be brought to management's attention immediately.

The Home Banking Service

It is difficult to imagine a better way to market retail banking services than to enable consumers to carry out their transactions in their own homes at their own convenience. Banks are making strides toward that goal as home banking proceeds in two directions: telephone and video. For a number of years numerous telephone bill paying systems have been in operation and video banking systems are being developed or even marketed. What is the state of those systems today? How about their development in the years ahead?

At year end 1981, in the United States alone, 168 Savings and Loan Associations (S&L) and 167 commercial banks were offering telephone bill payment (TBP) services, as were also 8 credit unions and 60 savings banks. Many banks consider TBP integral to their retail services, as indicated by the way they market it.

Though the number of transactions varied with the type of financial institutions, there were, on an industry average, 8470 calls a month and 2 transactions per call. But the range was wide: from 81 transactions at First Citizens Bank of Bozeman, Montana, to 271,475 transactions at Girard Bank of Philadelphia. Furthermore, a discrepancy was found between the number of customers who signed up for the service and the number who used it regularly.

The first videotex experiment took place in 1977 when Dow Jones introduced its News/Retrieval service. There were 5000 users of the service in December 1979, and as of late 1982 there were more than 33,000. Currently, 39 commercial videotex services are offered. Thirteen of the services are worldwide; 12 are in Europe; and 14 are in North America. Of the latter, 6 are bank experiments.

Although TBP is developed and growing in popularity, the greater potential and flexibility of video banking see to it that the latter will eventually be the preferred system. Some important decisions have to be made before home banking can take any giant steps toward its destiny, but no banker can afford to ignore the progress being made to take services into customers' homes. Chemical Bank's Pronto should be seen as a development effort following the track made by past experiences. In April 1982 it was put to trial use by over 200 customers, and it was subsequently introduced to residents of the New York City area on a fee basis. The technical configuration is uncomplicated: The consumer equipment consists of a personal computer, a telephone line, and a TV set. The only nonstandard equipment is a communications software package and a direct-access autodial modem. When a customer uses Pronto, a telephone call is placed to the bank's Tandem II nonstop computer.

We have spoken of two major areas of implementation: home banking (including account maintenance) and electronic message exchange. The former, we said, divides into major home banking functions: balance inquiry, funds transfer, checkbook register, electronic reports, and home budgeting. Let us now look at the component parts of each service.

Balance inquiry provides daily balances, summaries of all Pronto accounts, and details of activity affecting client accounts. A customer can obtain a detailed look at his balance any time he wants, whether the bank is open or closed, and he does not have to tie up a teller. Electronic reports are online, interactive account statements, summaries of Pronto and non-Pronto activity, and descriptions of all transactions such as checks and ATM-actuated Visa. These are available for current and preceding month consultation. The customer can add or delete offerings within the limits of the system and determine for himself how much data he wants to provide or receive.

Consumer-oriented electronic funds transfer includes shifting funds between accounts and making cash advances from revolving credit accounts—in general, more efficient money management. Through this online facility the bank can allow its customers as much freedom as it feels comfortable granting. By selecting eligible accounts and imposing the limits it wishes to impose, bank management can keep complete control over the service.

Bill paying is effected directly from accounts without writing checks, and the payments are charged directly against home budgets. They can be value-dated up to 90 days in advance, and the software also assures that recurrent bills can be paid automatically. The consumer can review, change, or cancel one or all payments he has directed until the payment is actually made. Payments to merchants can be effected by sending a check and a list and thereby transferring funds automatically or via an automatic clearing house (ACH). The customer can pay any merchant on the Pronto list of merchants and can add his own local merchant to the steadily updated standard list.

The electronic checkbook facility supported by Pronto records checks as written, monitors dates that checks clear, sees to it that checks are charged directly against home budgets, and assures that information concerning checks is stored for printed monthly statements. The financial institution using Pronto can offer electronic checkbook for any current account (DDA, NOW) and give its customers a first-class electronic reconciling service which is always prompt and up to date.

Home budgeting is addressed to the maintenance of complete financial records. It helps organize income, expenditures, and taxes and supports as many as five separate budgets. The software permits multiple entries, supports instant totals for any budget, and categorizes tax-deductible items as they occur. Home budgeting also offers monthly and year-end printed summaries. This is the beginning of a complete personal cash management system permitting account detail by checking available cash and line of credit and outlining the client input in terms of outstanding checks and payments to be made in the next 90 days (forecast).

In this man-information interaction supported by an intelligent workstation in each home and office, the client gives detail and gets it back through computer response. But both consumers and merchants can gain from the service, which in its first offering attracted, in New York City alone, 300 merchants who agreed to accept payments through Pronto. Account maintenance is, in fact, one of Pronto's most advanced features. Bank customers can actually run their own payment service. Paperwork is thereby greatly simplified, so the bank's staff does not have to assume extra burdens. Even when bank approval is needed, it can be granted electronically.

Flexibility is one of the system's characteristics. The ability to make changes is built into Pronto. Electronically, the customer can change his code name "handle" or personal identification code any time he feels an extra need for security. He can also add members of his household or request additional bank accounts. The system can, upon his request, grant him limited access, as for adding to his list of authorized merchant accounts. With equal ease it can add budget categories. The account maintenance programs permit him to be in control of his own banking matters, while it frees the bank's staff from tedious clerical work.

Customer access is based on a paging mechanism administered at the customer's site through menu selection. Since page routing is based on menu selection, access to it is, in the current offering, hierarchical, but the software developers plan to incorporate direct access as well. At the client's home both the Atari 400 and 800 can be used, but the 16K smaller model (Atari 400) is enough for now.

The first page the customer receives on his video unit is: "Welcome to Pronto." This is followed by a menu selection capability leading to a connection to central resources. There is, however, a relatively slow response time. The customer makes the following entries:

First entry. Household ID—everybody in the house uses the same ID.

Second entry. Citizen Bank handle. This can be alphanumeric or numeric—a nickname, for instance.

Third entry. Personal identification code (PIC).

Fourth entry. A payment order, which can be scheduled by the customer for any time within the next 90 days. This will typically include payment of $XX, to, say, Bloomingdale's from checking account XX, scheduled for XX/XX/XX. Description: November bill, charged to, say, the clothing budget.

The fourth entry demonstrates how the bank payment system connects to the family budget. The payments made through Pronto can be programmed to fit within the budget (in this case, clothing). To cancel payment, the consumer can

- Call the frame on scheduled payments
- Select the line of reference
- Call the page with the chosen reference
- Give a "cancel" command

Note in this connection that the customer assigns the PIC by himself to himself: neither his family nor the bank knows it; the PIC never appears on the screen. Hence, there is need for three levels of security. Put differently, Pronto was designed with several layers of security and predetermined limits of activity to minimize the bank's exposure. Management can, for instance, define transaction limits by amount, limit transactions by type, and use Pronto as an alert to potential overdraft situations.

Licensing and Implementation Goals

Chemical Bank's Home Banking System is licensed to other banks. The Crocker National Bank, of San Francisco, and the Florida National Bank, of Jacksonville, were the first licensees. At the time of writing, other U.S. banks had signed letters of intent and interest was growing. A dozen Japanese banks came to visit the system in the first three weeks of October 1982. The market goals set by Chemical Bank are a function of wider public ownership of the PC. At $1 per month per customer, to be paid by the licensee bank, 15 million microcomputer Home Banking System users in the United States would mean $15 million per month to Chemical Bank in earned fees. That would be a solid source of income. The home banking prospect is a good indicator of 1980s and 1990s income, which for any bank will largely be based on the delivery of information services.

To help the customer in his online communication with the database, Pronto maintains a list of corresponding banks and merchants. Through a four-digit identification, the customer can handle his bill paying and other obligations directly. If an outlet is not on the bank's list, the customer can add it. But Chemical Bank has been aggressively signing up merchants on the Pronto application. The following organizations were on the database in late 1982:

1. *Chemical Bank accounts.* Holiday Club, Installment Loan, MasterCard, Mortgage, Privilege Checking, Visa.

2. *Installment loans.* American Savings Bank, Bank Leumi, Bank of Commerce, Bowery Savings Bank, Central National Bank of New York, Chase Manhattan Bank, County Federal Savings & Loan, Emigrant Savings Bank, European American Bank, Jamaica Savings Bank, Long Island Trust, Manufacturers Hanover Trust, Marine Midland Bank, New York Bank for Savings, Union Dime Savings Bank.

3. *MasterCard.* Bank of New York, Citibank, Connecticut National Bank, European American Bank, Hartford National Bank, Manufacturers Hanover Trust, Marine Midland Bank, Midatlantic National Bank-Citizens Trust Company of New Jersey, Union Trust Co., United Jersey Bank.

4. *Mortgages.* Bank of Commerce, Bay Ridge Savings & Loan, Brooklyn Federal Savings Bank, Brooklyn Savings Bank, Central Savings Bank, Chase Manhattan Bank, Citibank, Columbia Savings & Loan, Dry Dock Savings Bank, Empire Savings Bank, First National State Bank of New Jersey, Franklin Savings Bank, Greater New York Savings Bank, Manhattan Savings Bank, Manufacturers Hanover Trust, Marine Midland Bank, National Bank of Westchester, Republic National Bank of New York, Roslyn Savings Bank, Scarsdale National Bank & Trust.

5. *Visa.* Bank of Commerce, Bankers Trust Co., Chase Manhattan Bank, Citibank, European American Bank, Irving Trust Company, Manufacturers Hanover Trust, Marine Midland Bank, Midatlantic National Bank-Citizens Trust Company of New Jersey, National Bank of North America, Scarsdale National Bank & Trust, Union Trust Co., United Jersey Bank.

6. *TV cable companies.* Long Island Cablevision, Manhattan Cable, Suburban Cablevision, Suffolk Cablevision, Teleprompter, Viacom Cablevision, Wometco Home Theatre.

7. *Clubs.* City Athletic Club, Doubles Club, Downtown Athletic Club, Knickerbocker Club, Metropolitan Club, New York Athletic Club, River Club, Union Club, University Club.

8. *Credit cards.* American Express, Diners Club, Carte Blanche.

9. *Department stores.* Abraham & Straus, Ann Taylor, B. Altman & Co, Bamberger's, Bergdorf Goodman, Bloomingdale's, Bonwit Teller, Brooks Brothers, Caldor's, Gimbels, Gucci, Hahne's, Henri Bendel, John Wanamaker, J.C. Penney's, Lerner Shops, Lord & Taylor, Macy's, Neiman Marcus, Ohrbach's, Paul Stuart, Plymouth Shops, Saks Fifth Avenue, Sears Roebuck & Co., Sterns, F.R. Tripler, Wallachs.

10. *Home heating.* Atlantic & Pacific Oil, Bergen Oil, Cibro Petroleum, Commander Fuel, Di El Corp. Oil, Franks Fuel, La Forgia Fuel Oil, Marathon Oil, Massapequa Oil, Meenan Oil - Suffolk County, Meenan Oil - Nassau County, Mobil Heating Oil, Murphy Fuel, Nassau Mutual Fuel, Nuzzi Fuel Oil, Paragon Oil, Perillo Bros. Fuel, Reliance Utilities, Rye Fuel & Supply, Schildwachter Oil, Shore Oil, State Utilities, Sunrise Oil, Super Fuel, Troiano Fuel Oil, Westchester Thermal, Weyant Oil Service.

11. *Insurance companies.* Aetna Life & Casualty, Allstate Insurance, Blue Cross-Blue Shield of New York, Commercial Travelers Mutual, Connecticut General, Equitable Life, Government Employees Insurance company (GEICO), Home Mutual, Liberty Mutual, Metropolitan Life, Mutual Benefit Life, New York Life, Phoenix Mutual, Prudential Life Insurance, State Farm Insurance, Teachers Insurance.

12. *Literature and news services.* Book-of-the-Month Club, *The New York Times*, The Literary Guild, *The Wall Street Journal.*

13. *Management companies.* AECOM Staff Housing, Armed Realty, Brookdale Village Housing, Century Builders, Chelsea Lane Apartments, Claredon Apts./What Cheer, Realty Corp., Clarendon Management, Clifton Builders, Dwelling Managers, Glenbriar, Israel Senior Citizens Development Fund Corp., Manhattan House, Nevada Towers Associates, Park City 3 & 4 Apts. Inc., Parkview House Co., Rhinelander Apartments, Rivercross Tenants Assoc., Roosevelt Hospital Staff Housing, Short Hills Gardens, Staten Builders, 372 Fifth Avenue Owners.

14. *Museums.* American Museum of Natural History, Guggenheim

Museum, Jewish Museum, Metropolitan Museum of Art, Museum of Modern Art.

15. *Oil and gas companies.* Amoco Oil Co., Atlantic Richfield Oil Co., Chevron Oil Co., Cities Service Oil Co., Exxon Oil Co., Getty Oil Co., Gulf Oil Co., Mobil Oil Co., Phillips Petroleum, Shell Oil Co., Sunoco, Texaco Oil Co.

16. *Schools.* Adelphi Univeristy, Chapin School, Fairleigh Dickinson, Hunter College, Katherine Gibbs School, Pace University, Trinity School.

17. *Specialty foods.* Balducci's, Mr. Lobel & Sons, Zabar's.

18. *Specialty stores.* Cartier Inc., Harry Winston, Tiffany & Co., W&J Sloane.

19. *Transportation.* Avis Car Leasing Co., Briggs Leasing Co., Conrail Mail-Tik, Eastern Airlines, Fugazy Auto Leasing Co., Lease Mobile, Long Island Railroad, National Car Rental Co.

20. *Utilities.* Albertson Water District, Bethpage Water District, Bridgeport Hydraulic, Brooklyn Union Gas, Carle Place Water District, Con Edison, Farmingdale Village Water District, Franklin Square Water District, Freeport Village Electric, Freeport Village Water District, Garden City Park Water District, Glen Cove Water District, Greenlawn Water District, Hampton Bays Water District, Hartford Electric Light, Hicksville Water District, Jericho Water District, LILCO, Locust Valley Water District, New Jersey Bell, New York Telephone, Oyster Bay Water District, Riverhead Water District, Roslyn Water District, Southern New England Telephone, Suffolk County Water District.

The Pronto database is run on Tandem computers. There is little information on the cassette that fits the Atari PC. Most of what is used in man-information communication is on Tandem.

To market the service effectively, Chemical sponsored five Pronto days. One was addressed to the financial institutions already using BankLink, the second to the VP Marketing of the top American banks, a third to the smaller banks, a fourth to merchandising companies, a fifth to the consumer (repeated several times). As with every seriously studied new product launching, management asked itself the questions:

■ Who are our most likely Pronto customers?

■ How do we reach them?

■ What is the most effective way to demonstrate Pronto?

■ How many customers should we start with?

In planning its marketing moves, management also considered the user manuals, demonstration sets, and promotion material to be important features of the Pronto package. Brochures, slides, advertising, and sales suggestions were developed to help both the Chemical Bank and the financial institutions which had adopted Pronto.

Believing that the concept of licensing new banking services will shape the growth of banks for years to come, Chemical is committed to continuing the development of its Pronto service and make it available to other banks on a licensing basis. That makes both economic and technological sense, and management hopes that Pronto will set an industry standard. Licensing goes through stages. First comes pilot licensing. The advice Chemical gives the licensee bank is to take a random sample cross-cutting 200 customers and not taking, for example, only customers who are computer people. Solving the interconnect problems at the sample level will make the setup easy. To help explain to the customers in the sample how to work with the service, Chemical and its licensees depend on the Pronto user manual. It illustrates everything from how to open the Atari box to how to set up and use the service.

Second, there is a pricing schedule for the experiment. In the United States, the fee is $200,000. Of this budget, 25 percent goes to Chemical Bank for its assistance in the pilot licensing experience, another 25 percent pays implementation and maintenance of the Tandem computer (if the central computer facilities in New Jersey are used by the licensee), and the remaining 50 percent is the cost of purchasing 200 Atari computers and Hayes modems.

After the pilot phase, and provided the system is adopted, full license requires $75,000 (implementation II) plus $10,000 and $1 per month per customer as royalties. These fees cover a full range of services:

1. Home banking

2. Electronic mail

3. Account maintenance

4. Investment advisory

5. Teleshopping

As stated in the introduction, all participating banks and all customers of participating banks have electronic mail facilities. Typically they will pay local call costs for coast-to-coast teletex. (A Tandem-to-Tandem connection is made through Tymnet.)

Although Atari is the PC currently used, Chemical plans hardware expansion. Said a senior executive: "We want every PC maker to be able to say 'we are Pronto compatible.' Hence, we will offer Apple, IBM PC, Commodore VIC 20, and Texas Instruments. We plan to handle every

PC with a good share of the market." (There are, however, some technical problems. TRS 80, for instance, has 32 columns versus 40 now supported.)

Online Wholesale Banking

The development of home banking as a retail, consumer-oriented system has been largely based on prior applications of online capabilities of computers and communications in financial institutions. Typically, these implementations were used in wholesale banking. Two precise Chemical Bank examples are ChemLink and BankLink. In operation since the mid-1970s, ChemLink is Chemical Bank's global computer-based cash management system designed to assist corporations in optimizing cash management effectiveness. It provides a secure and reliable means of monitoring, analyzing, and moving funds on a domestic or global scale, and it was designed for treasurers, cash managers, and credit managers in domestic and multinational corporations, financial institutions, and government agencies. In terms of service, ChemLink will:

1. *Access* accurate and timely information to summarize a company's domestic and international cash positions

2. *Analyze* the cash position against target balances

3. *Invest* available funds

4. *Initiate* fund transfers, foreign exchange transactions, securities clearance, and settlement around the world

5. *Receive* continuous information on the customer company's worldwide investment portfolio and foreign exchange rates

ChemLink uses Mark III/GEIS to access the global database 24 h/day, 7 days/week. Security is supported through user number ID.

Through a third-party computer network and a series of security gates, ChemLink ensures confidentiality. When a client company joins, it is issued a private code so that only authorized persons in the firm have access to confidential records. Not even Chemical Bank can gain access to the accounts of its own customers at other inputting banks. To the client, ChemLink can provide account information on other banks worldwide. The balance report is on a previous day basis. A current account module works for the current date and is subject to verification the same day.

ChemLink has been designed as a comprehensive and flexible system which can be personalized to fit a company's needs. The system has more than 1500 banks around the world reporting information. This input base makes it the largest interactive cash management network in existence. A range of reports are available:

1. *Balance.* Provides one consolidated report covering a company's worldwide cash position, including gross and collected balances and debit and credit totals.

2. *Balance history.* Supports up to 2 months history of daily balance information.

3. *Debit and credit.* Reflects daily account activity.

4. *Lockbox.* Offers same-day monitoring system for all lockbox reporting.

5. *Target balance projection.* Monitors month and year-to-date performance for your account at any bank. Target changes can be easily implemented.

6. *Custom reports.* Can be tailored on request to meet specific requirements of the client firm.

For international financial data, ChemLink interfaces with the CEDEL Eurobond clearing system based in Luxembourg. Through it, it monitors worldwide Euro-security transactions in various currencies. Customer reports include transaction settlement status and portfolio positions, and the user is provided with:

- *Foreign exchange rates.* Spot rates for 37 countries updated three times each day and expressed in 18 base currencies. Forward rates for over two dozen countries are provided.

- *Short-term rate.* Covers daily market rates including Chemical's prime rate, federal funds, T-bills, gold fixings, and Eurodollar bid rates.

BankLink was an outgrowth of ChemLink. Fundamentally, ChemLink is licensed as BankLink to leading banks around the world. It enables client banks to offer their customers the same services that Chemical offers. Through its member banks, BankLink serves a wide spectrum of clients; they include some of the world's largest companies but also smaller businesses, schools, hospitals, local and state governments. Some 50 banks are licensees.

Like ChemLink, BankLink is an information delivery mechanism which goes well beyond balance reporting. It offers:

- Cash movement

- Forecasting

- Cash management

- Analytical studies

Chemical Bank sees licensing to other banks not only as a good policy for service income but also as a vital step toward increasing the pool of ideas in development. Licensing is instrumental in helping the system evolve over time. The ChemLink and BankLink systems assure transactional information for customers through the GEIS service accessible through Terminette (GE), Silent 700 (TI), telex, touch-tone phones, and the projected PC ports. The software is modular:

- Each module performs a different function.

- A basic package of modules is given to the member bank to fulfill the cash management requirements of its clients.

The member bank has the option of selecting additional modules that move funds, issue market rate reports, gather additional information for transaction monitoring and account analysis, and manage securities trading activity. It can choose any or all of the modules and pay only for the functions being used. Chemical Bank's philosophy along this line is that the market for computerized cash management is large and is expanding rapidly. Increasingly, financial officers and businessmen depend on financial institutions that can simplify and standardize the daily information flow necessary to mobilize and utilize cash.

Chemical's marketing policy sees to it that the licensing arrangement assures minimum risk and maximum efficiency to a bank when entering the field. By sharing in BankLink, the financial institution gets a hold on computerized cash management. The following is a brief description of the supported modules:

1. *Basic cash management.* This is multibank balance reporting with status information, balance history reporting, previous day debit reporting, previous day credits, transaction code listing, depository transfer checking, lockbox reporting, and audio-lockbox reporting.

2. *Movement of funds.* This module handles preprogrammed funds transfer, free-form funds transfer, and foreign exchange transactions.

3. *Additional transaction monitoring and account analysis.* Its object is current-day debit and credit reporting, 2 days previous debit and credit, target balance projection, and special reports.

4. *Market reports.* Mainly foreign exchange rate and short-term rate reporting.

5. *Security activity.* A security activity report and securities order entry.

Balance format information includes account number, posted date, gross and collected balance, one- and two-day float, total credit and debit amounts, and number of credits and debits. The user can choose from 20 formats, and further customization for specific customers is available. Reports can show subtotals either by bank or by selected groups of accounts in several banks of subsidiaries.

Status format information includes account number and last posted date of balance information on file. It saves valuable time and expense by determining the availability of account information before using the balance, credit, and debit modules.

A bank can offer its customers audio-balance reporting (ABR) via BankLink. This feature enables the customers to call for balance information with touch-tone telephone or rotary telephone with touch-tone adapter. Because it is used without a terminal, ABR is ideal for non-computer-based operations. The following items are available for each account:

- Ledger and collected balance
- One- and two-day float
- Total debit and credit amounts
- Number of debits and credits

In addition to providing daily balances, BankLink stores multibank information for over 2 months' activity. A bank's customer may request a specific date or a range of dates for up to 65 days of information in full detail or in totals and averages. This module provides, by account, comprehensive daily reports of:

- Gross and collected balances
- One- and two-day float figures
- Total debit and credit amounts

An average is given for each category displayed, and totals are supplied for debit and credit categories. To support a customer's daily account reconciliation, BankLink assures a complete listing of the previous day's transactions through the detailed debit and credit report modules. Formats are designed to be easily understood; each report displays item-by-item descriptions including the source or beneficiary of each transaction. Debit and credit modules enable the customer to:

- Monitor significant individual items as they are posted
- Track items of particular concern
- Track items for error correction
- Take advantage of investment opportunities
- Provide an audit trail for accounting purposes

By using the depository transfer checking (DTC), the client bank can move funds into its customer's main concentration account, monitor current-day deposit activity at decentralized locations, accumulate funds to a prespecified total, and set a maximum DTC amount for each location. The customer also gets centralized control to override standing instructions and change deposit totals at any of the deposit locations, obtain hardcopy reports of individual deposit amounts made daily at each location, and identify nonreporting locations and take corrective action.

Coming Developments

The wholesale online services to which we refer have been implemented since the mid-1970s. They are, however, subject to new developments, and a specific example is the use of personal computers and communicating databases. Economics definitely favors PC implementation as opposed to timesharing. An approximate $3 per transaction charge on timesharing is not a cost-effective solution and tends to limit system use. Microcomputers help alter the economic considerations radically. Chemical Bank, for instance, believes that PC use is the next generation of online wholesale banking. This will make feasible the operation of local databases with downline loading and upline dumping capabilities. Indeed, management is committed to move in that direction.

The integration into the online system of gateways and communicating databases will foster the evolution of a new transaction-oriented, microprocessor-based system with much higher sophistication. This in turn will lead to a new relation, on a daily basis, between the client firm and the bank going well beyond current-day report capabilities. To appreciate the extent of those developments, let us first look at the current implementation from which the new system will start.

- Fifty banks are already on BankLink.

- Another 30 banks are not on BankLink but have high-speed lines connecting them to Chemical.

- Some 1800 other banks can reach into BankLink via terminal, touch-tone phone, telex, and so on.

Some banks use the Swift-type format. Though BankLink has its own format, there is a translation to Swift for both incoming and outgoing messages. Local intelligence (through PCs and minis) and communicating databases will further enhance the automated instruction with which the customer communicates online by means of a data communication rather than a voice facility. The projected system approach is shown in Fig. 16-1.

Each member bank operating on this system has its catalog through which it can perform any desired data exchange. The corporate client can receive information from accounts it has with other banks participating in the system, and routing capabilities also are supported. The participating bank can input money rates and update them for customer reports. The customer firm can ask for a specific report; hence, the system is informative, executional, and analytical. The report can be multibank, multicurrency, and multiformat. The customer can choose subtotals by bank, account, currency, and division. With a PC installed at its site, the customer firm can further elaborate on a report to obtain decision-oriented information, graphical presentation, and the use of color for the exceptions.

Just the same, local intelligence can extend the capabilities of lockbox reporting, which currently provides a same-day monitoring system for the customer's lockbox network. Participating banks input information directly into the BankLink system regarding:

- Account number

- Box number

- Deposit date

- Deposit time—a.m. or p.m.

- Deposit amount for each lockbox

Multiple display formats are included. They provide input of available

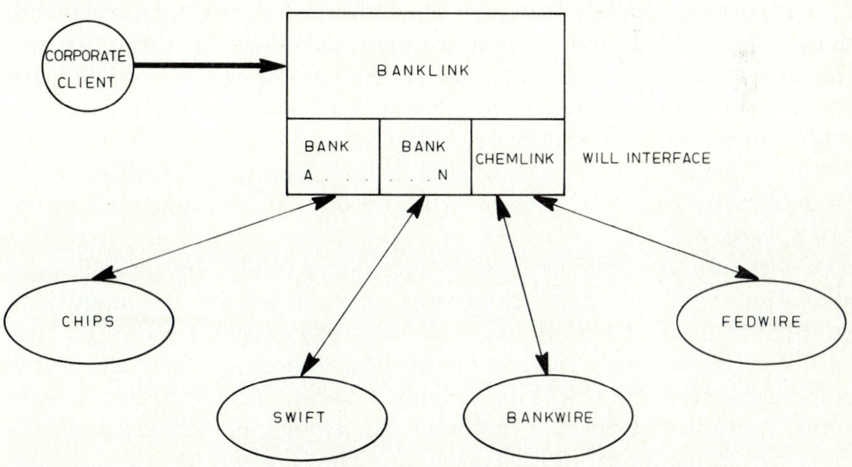

Figure 16-1 An automated communications link can handle online a myriad of interfaces with corresponding banks and national or international message-switching networks.

funds and float totals if customer firms request that information. Benefits include a timely, accurate consolidated report and convenient tracking of daily receipts by location.

Locally installed PCs will enhance, through graphics and color, services currently supported by audio lockbox reporting. This feature enables customer firms to monitor their lockbox network via touch-tone telephone or rotary phone with touch-tone adapter, and the unsophisticated end user will thereby have entry to the field. Personal computers at the end user site will greatly enhance funds transfer capabilities. Customer firms require an accurate, secure, and uncomplicated system for moving funds.

BankLink already supports two easy-to-use funds transfer methods:

1. Preprogrammed funds transfer

2. Free-form funds transfer

Both methods make it feasible to initiate transfers and generate hardcopy confirmation of all instructions. However, funds transfer based on local intelligence can significantly benefit any firm initiating repetitive payments including the operation of a local database containing the actual time the order was entered or received by the bank, the time when the bank confirmed the transfer, unique identification numbers, and local availability of information elements to enhance auditing, reporting, and control.

Additional activities can be promoted along the line of transaction monitoring and account analysis. BankLink's current-day debit and credit reporting already provides an updated list of items processed, which enables the customer to monitor current-day activities and track items of particular interest. Available reports show a detailed description of each debit or credit that has been posted to the customer's account on an individual item-by-item basis. The source or beneficiary of each transaction also is indicated. Local intelligence will improve this facility by supporting an internal accounting management information system.

The same thing is true of possible extensions to the foreign exchange transactions (FXT) module. Its current aim is to benefit customer firms who initiate nonnegotiated and repetitive payments in a foreign currency. The customer can initiate prespecified buy or sell instructions for any foreign currency traded by a member bank. With prior acceptance of current trading rates, the customer can accurately initiate and confirm instructions for foreign payments on the BankLink terminal. The firm can also inquire about the state of previously entered transactions. Reports are retained for 2 months, and a monthly recap and summary also is available.

BankLink assures timely information needed to stay abreast of rapidly fluctuating foreign exchange rates. Spot rates for 37 countries are included in its foreign exchange rate summary. They can be reported in 18 different base currencies, and standard reports can be established for those currencies relevant to the customer firms' immediate needs. Local intelligence will further extend the currently supported facility to access and manage securities activity in custody accounts. The daily reports now available, which detail prior-day securities settlements such as purchases, sales, and income received, can be shaped through the PC in any format the end user requires, including graphics.

Typically, such reports indicate standard security description, units of face value, and the dollar amount for each transaction, with data grouped into the following categories:

- Sales and principal receipts

- Purchases and principal disbursements

- Income receipts

- Income disbursements

- Memo receipts

- Memo disbursements

Early access to hard- and softcopy confirmation of settlements, net cash position, rapid daily transaction reconcilement, and detailed interest and dividends credit are among the current operations which can be better shaped through computer power at the end user site. That is equally true of the special report modules supported by BankLink. Their aim is to give the customer firm broader capabilities in the area of service differentiation. By using such modules, the member bank can transmit to a customer any report which it produces and thereby eliminate the lead time often associated with tailoring a new module for each procedural solution.

As international licensees (banks) join the system, the software will be developed to make the exchanged information more meaningful to the end user. That is true of both language and the accounting practices involved (value dating, and so on). Local intelligence makes it feasible to distribute functions which, though useful to some firms, are not universally required and let the mainframe fill the role of a big message switching center and database machine.

17

The Pronto System Software

You only win if you control the game.
OLD PROVERB

The best way to look at the system structure of Pronto is as a series of successive releases each improving upon the performance of its predecessor. Improvements go all the way from central computer systems support to communications channels, protocols, and personal computers at the end user's site.

Figure 17-1 contrasts the 1982 and 1983 releases. In both, the screen drivers are written in Scobol (a Tandem screen Cobol for interactive applications). But the 1982 version uses IBM's 3270 protocol, which has been dropped in the new release. (The 3270 protocol works in three passes, and that is too slow for interactive applications. The first pass keeps positioning the cursor; the second pass contributes the attributes; and the third pass paints the data in. That is very inefficient. The new solution sweeps the screen as videotex does: left to right, up to down. This results in a saving of 55 percent on straight transmission over the 3270 protocol.)

Major Improvements

Let us look at the response time characteristics of the Pronto system. It now takes 2 to 3 s plus a built-in 2-s delay to assure the message is complete. The new release will do away with the delay and will therefore be faster.

The second major change made with the new release is the substitution of a front-end processor for the Datastream unit employed in the original system. With the first release, the Tandem computer talked to Datastream and Datastream talked to the Atari PC. All security and protection was on the Datastream. The substitution brings with it five important features. First, the central machine will assume the security and protection responsibility. Second, the FEP will be interfacing to X.25 packet-switching value-added networks. Though the latter is a requirement, it constitutes a handy enhancement where X.25 VAN are implemented and opens the way to the implementation of communicating databases. Third, as to software, there will be a switch to Tandem's routines for online checkpointing. With it will be installed an automatic operating subsystem and a tape library facility.

The fourth major improvement is at the home banking (consumer) level; it is the capability of keeping eight pages on the Atari 400/800 PC. This permits local recall and, as a result, the saving of some 80 to 90 percent on transmission costs. The fifth improvement concerns the ability to handle the Atari as an intelligent machine. All home banking work will be done locally on the PC until the customer is ready to send. All messages are protected during sending and receiving by longitudinal redundancy check (LRC, Tandem's protocol). Not that the home user works start/stop full duplex at 300 baud. The system can support 1.2 kbps, but that calls for special jamming of the channel.

As with videotex, the pages in the database are accessed through menu selection. Lines and characters are manipulated by cursor positioning on Atari. Also on Atari are the security measures. Direct access

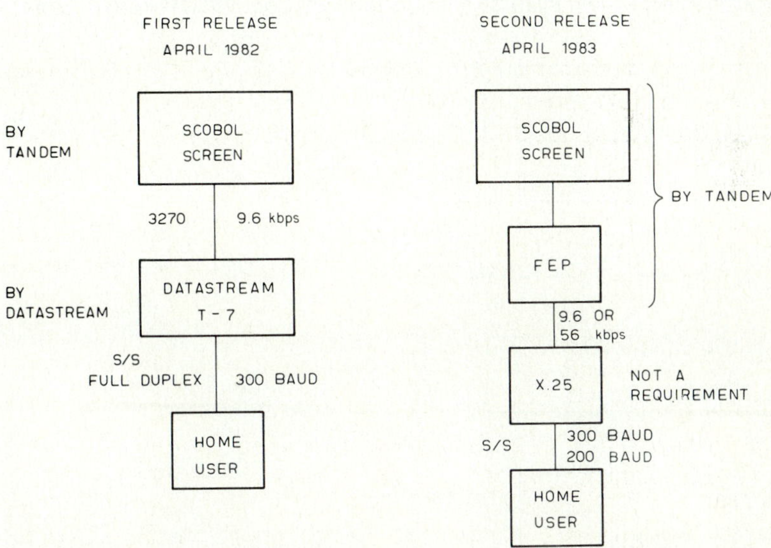

Figure 17-1 The second release of Pronto software presents significant system improvements over the first. A third release will incorporate the PLP standard.

to pages is not yet available, but the need is recognized and the facility will be included in a third release. Other additions to the two Atari machines have actively been considered, and some are in course of implementation. That includes a PC with enlarged local storage capabilities (microfiles). The goal is to eventually let home users work with floppy and hard disks.

The PCs other than the Atari announced as Pronto-supported are IBM and Apple. All PCs supported and to be supported employ the same program facilities. The software is written in C under UNIX on a dedicated VAX. System design is highly modular, and it follows a layered approach. As Fig. 17-2 demonstrates, new applications can be added to the system through an update of the *server* interface: Its task is to get the account information and access the file. The screen management (Scobol) and front-end routines remain unaltered. Attachment to existing subsystems, whether realtime or batch, is simple and straightforward. The interface to other, older batch-oriented procedures is through the access file. For realtime subsystems, the communications line serves as the link.

Toward an Electronic Teller

Three other issues are very important and must be brought into correct perspective:

1. *The structure of the Pronto system is based on the ISO/OSI model.*

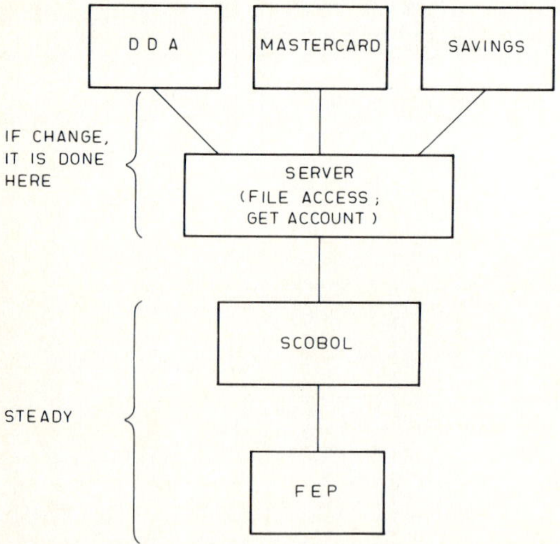

Figure 17-2 Connecting the consumer to the branch office level. System design sees to it that new applications can be added through an update of the server interface.

International standards are observed not only in layering but all the way to the graphics. The graphics standard is presentation level protocol (PLP), but the way has been kept open for integration with the CEPT standard.

2. *System design foresees the capability of attaching at the front end an impressive array of services to the business community and the bank's own branches—over and above home banking.* As Fig. 17-3 shows, public terminals, POS, and the business information system (BIS) will be interfaced to Pronto in 1983. The BIS is addressed to corporate cash managers and the service is provided through file transfers and audio response. To a fair extent this is a substitute for the BankLink and ChemLink setup, which was originated in 1975 and is still in operation.

The "public terminals" will be Pronto stations installed in shopping centers and Chemical Bank's own branch offices. These are polyvalent three-station machines; two stations are ATM, and the third is a touch-sensitive screen similar to the AT&T installation at Epcot Center. The latter is part of the developing *electronic teller* concept. The marketing end of it uses videotex (optical recording) to inform the client, and the touch screen works on menu selection. It permits signature verification, and the whole block is controlled by an Apple II and costs less than one standard ATM unit. This solution is open to local area network (LAN) implementation.

Chemical Bank plans to install the electronic teller in all its 260

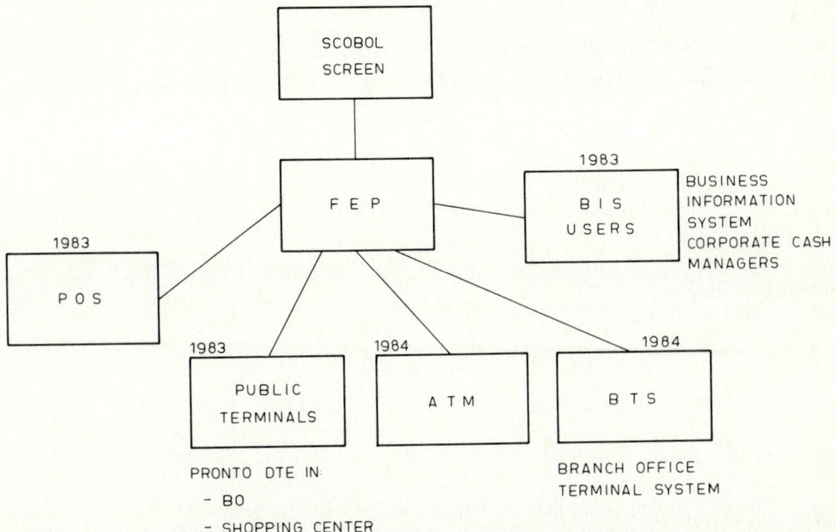

Figure 17-3 Project growth of the Pronto system to cover every aspect of the retail banking market and also some wholesale applications.

branch offices and eventually in vestibules. Studies have shown that it can outperform the live teller, who usually performs 225 transactions per 6-h day. Though the simpler transactions will be undertaken first, in a typical branch office 84 percent of transactions are demand deposit account (DDA) and savings accounts. "It is a matter of marketing the new services," said a senior executive. "But even an Atari can do the job the teller now does. He is just not doing that much." This development makes the attachment to Pronto of the ATM currently in operation feasible in 1984. Most importantly, the same statement is valid for the branch office terminals.

On the platform area, for consulting purposes, there will be a color graphics display and suitable standard products for the officers, together with a print server when needed. The overall object is personalized service to the customer.

3. *A multilingual capability for the interactive screens* to be achieved through the attachment of a *screen library* to the front-end processor, was scheduled for early 1984. To see how the multilingual system works, we must remember that:

- All intelligence resides in the Scobol screen.

- The front end processor (FEP) is used for file transfer.

- The screen library capitalizes on the parametric flexibility built into the screen structure.

- The end user can manipulate the macro without necessarily knowing how the system works.

By changing the macro, the screen is reset automatically. By editing the screen, the macros are automatically affected. The menus are dy-

SCREEN USAGE	MACRO
CHEMICAL BANK HB	# MACRO # 1 #
1. HOME BANKING	* P 1 *
2. ELECTRONIC MAIL	* P 2 *
3. ACCOUNT MAINTENANCE	* P 3 *
	# MACRO # 2
PRESS BUTTON TO CONTINUE	#

Figure 17-4 Screen library application. The use of the PLP protocol will permit highly flexible screen editing to be implemented with ease at the home banking PC.

namic. The end user can edit out lines, and he can select those options he wants to see. Figure 17-4 shows a screen-macro correspondence.

Although the software has been developed by Chemical Bank, the macro picture description instruction (PDI) is by AT&T/PLP. This presentation level protocol helps in implementing different language versions. Instruction writing is very simple:

Macro 3 = P1 + P2 + P3

Screen = Macro 1 + M 2 + M 3

Instead of storing screens on the terminal, we can store macros. A macro can be color, graphics, text, or data. It can contain one to an infinite number of characters, but three characters are needed to define it. The PLP protocol makes the implementation of Pronto significantly different from that of Prestel.

Let us add that, in the first release of Pronto, the macro storage is on Tandem. The implementation of new and more powerful machines will handle the macros locally (at the client site), but the understructure will not be altered. The implementation of PLP in actuating the file transfer protocol pays handsome dividends.

The careful, far-reaching design by the Chemical Bank staff sees to it that the system will run PLP at the terminal also, and not just between the central processor and the PC. Note, however, that the solution is an objective of the new release. Plans called for it to be the subject of experimentation for home banking in 1984 and its possible business implementation in late 1983.

The fact that the PC is capable of PLP implementation streamlines the downline operations, makes the consumer-to-computer interaction user-friendly, and observes international standards. As Fig. 17-5 dem-

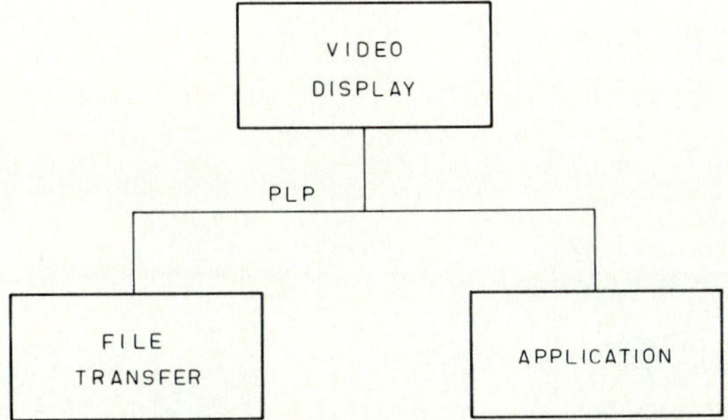

Figure 17-5 Through PLP implementation with the PC it will be possible to integrate at the workstation level the user-oriented applications programs with file transfer capabilities.

onstrates, it also helps integrate, at the video display, file transfer and the result of processing associated with a specific application. This has been a very good choice by the system designers.

Touch-Sensitive Screens

Touch-sensitive videoscreens provide a simple and relatively effective way to communicate with computers. Rather than push buttons, switches, or keyboards, the user touches a computer-generated display on a terminal screen. The screen must be equipped with a touch-sensitive device. Through the device the computer determines the spot on the screen which has been touched. It then takes the action for which it has been programmed.

Typically, a touch-sensitive terminal uses a pattern of closely spaced horizontal and vertical infrared beams which pass very close to the face of the display (Fig. 17-6). As a finger touches the screen, the beams corresponding to the spot touched are interrupted. That permits the computer to identify the exact coordinates of the spot touched. To put it differently, the key to touch sensing is the matrix of infrared photodiode transmitters and photodetector arrays that frame the video terminal screen. Upon sensing the interruption of the infrared rays by the user's finger, the photodiodes send the videoscreen coordinates to the computer for processing.

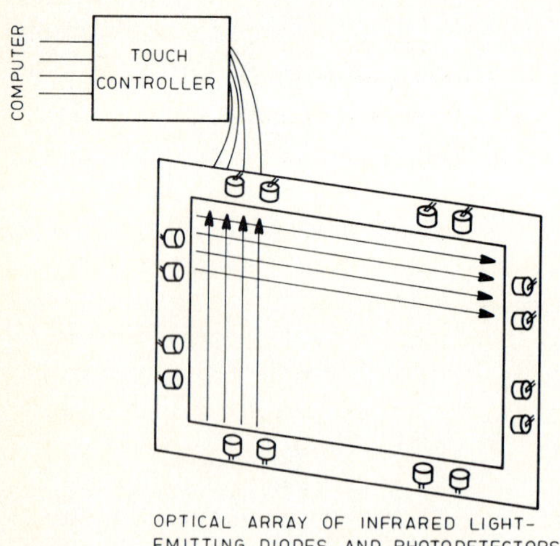

Figure 17-6 A touch-sensitive screen uses a pattern of closely spaced horizontal and vertical infrared beams near the face of the display.

A touch-sensitive terminal has four major components:

1. The color video monitor
2. An optical array of infrared-light-emitting diodes and photodetectors
3. A protective plastic cover
4. A microprocessor touch controller linking the terminal to the computer

In one of the available units, the matrix is composed of 64 infrared light-emitting diodes positioned at the bottom and 48 along the left edge of the optical array. An equal number of photodetectors are located opposite the diodes along the top and right edges. To determine accurately when and where the user is touching the screen, the microprocessor in the touch controller monitors each horizontal and vertical beam individually in sequence. All beam positions are scanned about every $\frac{1}{15}$ s.

The processor senses the amount of light at the photodetector just before the opposite light-emitting diode is turned on and stores that information. Then it tunes the infrared diode opposite the detector, senses any changes in light level, and finally compares the two light levels. The photodiode touch detectors have a resolution of about ⅛ in. If the scanning procedure detects a significant increase in light, the controller has the evidence that a finger has not blocked the light at that position. No significant increase in light indicates that the light beam has been interrupted. The controller microprocessor computes the exact coordinates at which the beam has been interrupted and sends that information to the computer for processing. The coordinate information is used to determine the programmed sequence to be displayed on the color video monitor.

Application of the touch-sensing technology can range from banking systems to any type of information provision, including tourist information. In one implementation, the video information provided in response to queries comes from optical video disk players. Each video disk contains 54,000 separate video color frames. In addition to a minicomputer, the terminal controller contains special graphics display and musical sound synthesizers.

A block diagram of the computer and communications facility is given in Fig. 17-7. The synthesizer provides audio responses, and also background music, to touch requests. If the user asks to speak to an attendant, the terminal controller requests the central control function to perform the video, voice, and data switching and so complete the two-way video connection.

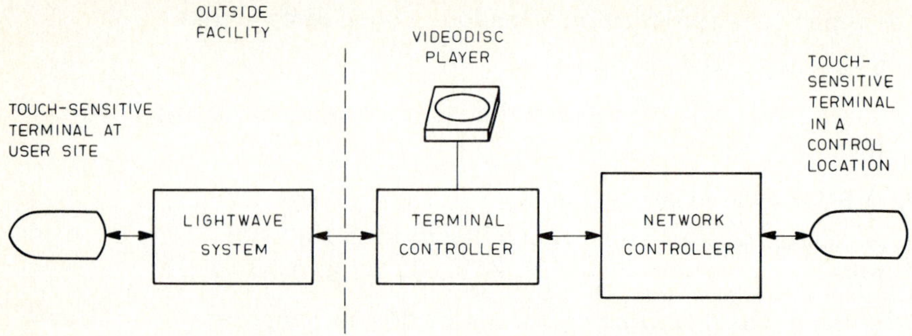

Figure 17-7 An integrated microcomputer-supported system with video disks and touch-sensitive terminals can be a key feature of merchandising terminals from banking to retailers and the entertainment industry.

The application under discussion was designed by the Bell Telephone Laboratories and is supported by a specially developed high-level computer language for programming the video disks. The programs are referred to as *scripts*. A video dictionary relates alphanumeric names to frame numbers on the video disk, and a flowchart debugging capability enables the script writer to correlate the video dictionary with the video disk.

Bell Laboratories have also been engineering a language to permit programming directly on the television screen, which will make it possible for the script writer to *finger paint* sections of the screen that are to be sensitive to the user's touch. Interactive software will allow the script writer to create the whole set of alphanumeric characters and symbols used in the graphic overlays. The operating system is Unix.

The WorldKey Information Service at Epcot Center, Florida, consists of 29 guest touch-sensitive terminals with up to eight terminals connected to a guest terminal controller. The latter supplies the audio and visual information about the center. Special requests are answered by attendants at seven touch-sensitive terminals. The network controller switches video and audio communications between a guest and an attendant and also updates information stored in the guest and attendant terminal controller. All video, audio, and data information is transmitted over a lightwave local area network.

The Videotex Experience of German Banks

"We must feed our opportunities and starve our problems." PETER DRUCKER

The 1982-83 experience with Bildschirmtext (videotex) in Germany has documented both the feasibility and ease of paying bills, doing banking, and generally handling credit and financial transactions over the network used for voice communication. In other implementations, particularly in the United States, the cable that delivers TV programs is used.

The technology is available, but that is not necessarily true of the procedural matter and the users' preparation to deal with the facilities we can offer. We must translate our experience in helping end users into the information services that will be available over cable and other two-way media. Furthermore, some crucial issues must be settled fairly soon before major problems emerge.

One key issue is the standard to be used for communications. Currently two standards are used:

1. The North American PLPS

2. The European standard, CEPT

We do know how difficult it is to change standards once they are established, and the decision made in an international conference in Geneva, in February 1983, aimed to answer the need for coordination through the implementation of a mechanism for switching between different sets of standards.

A standard for videotex services will mean better communications and a clearer presentation. The new European standard looks very much like a 50/50 compromise between the alphageometric and the alphamosaic solutions, and the standard adopted in North America is a fully alphageometric solution with an advanced presentation level protocol. Although differences can be made transparent through gateways, it would have been more rational to have one standard for the whole system worldwide. That, as we stated, was sought in Geneva by the international committee (CEPT, PLP).

A second major issue is the vehicle that will be used for information communication. Will it be coaxial cable, optical fibers, voice-grade telephone lines, or satellite? Each can be argued for. Coaxial Cable has an established technology and supports high-resolution screens. Optical fibers provide much higher capacities, but they do not yet allow multitapping. Telephone lines are available in more than 98 percent of postindustrial society (the United States for example), and there is little or no installation problem. Satellite communication for home services is still in its infancy, but the broadcasting capacity is enormous. Although it is too early to predict which vehicle will triumph, the decision made could drastically affect future development.

A third area in need of careful study is the type of terminal to be used. Will it be a classical computer terminal, a typical television set, or a personal-computer-based facility? What about the specially equipped telephones being used by some financial and industrial organizations? The willingness of consumers to buy special terminals specifically for home banking and shopping-at-home transactions is a definite consideration in this connection.

Bildschirmtext Implementation

In the implementation plan of Bildschirmtext (videotex) on a nationwide basis in Germany, three basic steps were foreseen: First came the decision on standardization. At the same time, IBM, the main contractor for the network, was expected to present its system. Parenthetically, IBM developed this network on a fully centralized basis with the mainframes to be installed in Ulm and 11 regional centers based on a Series/1 computer. The software was not yet ready at the time of writing, however.

As the original plan stood in September 1983, the first implementation of the new system was to be launched with the occasion of the Telecommunications Exhibition (Funkausstellung) in Berlin. The eleven regional centers were projected to become operational by April 1984. Many IPs based their plans on those premises. KAUFHOF, the merchandising and travel organization, is an example. After development of its Bildschirmtext use package, the first implementation took place in September 1982. The connection to the Datex-P net of the German

Post Office supports a transmission speed of 9.6 kbps and distribution to 10 virtual connections which can be used simultaneously.

In spite of the tentative character of this solution (because of the Bildschirmtext field trials), we can assume use of the current system until the end of 1984. The user organization considers its investments are justified, particularly because of:

1. The accumulation of experience with this new service
2. The marked improvement in the information flow which management considers to be a strengthening of competitiveness

The reference to the Datex-P link must be further explained. As Fig. 18-1 demonstrates, the German Post Office plans to link the videotex network with the Datex-P (packet switching network). Bridges will be established node-to-node at the regional videotex centers on the suppo-

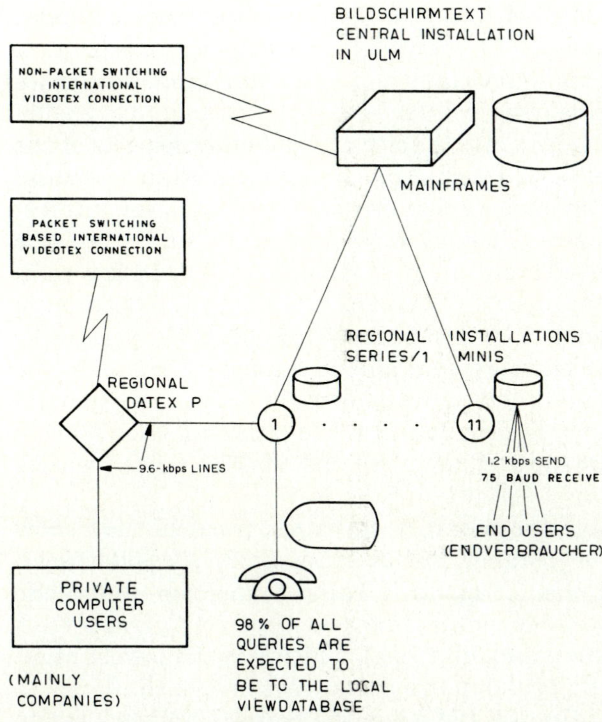

Figure 18-1 The German Post Office (Bundespost) plans a dual system of videotex communication: Datex-P for database exchange and a query facility for private users. Bridges will assure internetworking.

sition that the Series/1 computers can manage the load. That they can is not at all evident. The Series/1 machine was launched in 1977. Today there are PCs much more powerful than this aged, never successful IBM mini.

Personal computers will one day be installed in small businesses and households, which will use the Bundespost facilities as a switch. The results will be quite taxing for the aged Series/1 equipment now installed, because the regional computing centers must work nonstop 24 h/day, 7 days/week under heavy load. The old minis cannot meet that requirement. And as if the poor choice of switching equipment were not enough, the $20 million contract for the first stage of the Bildschirmtext system ran into software troubles—in spite of IBM's software capabilities. As should have been expected, the complexity of the system, which will serve 1 million customers with 400,000 online terminals by 1986, is far more than IBM originally bargained for.

IBM was reluctant to discuss problems connected with the development of the system, but Bundespost admitted that it would have to delay some services by as much as a year because of software problems. At the same time, from IBM's viewpoint, development costs have mounted and some experts believe they may wind up at more than twice the amount of the original bid.

The implementation delays due to software and other aspects of the nationwide videotex network are that much more regrettable because the Berlin and Düsseldorf experiences demonstrate that user interest is growing from year to year. Though the system is still under test, the statistics now available substantiate that statement. Within one year, the number of input pages grew in Berlin by more than 120,000 and in Düsseldorf by more than 100,000. In both areas, about 350 percent as many Bildschirmtext pages were available in January 1983 as in the beginning of 1981. Still greater was the increase in interactive consumer contact. From January 1982 to January 1983, the calls doubled. In Berlin, from January 1981 to January 1983, the number grew sevenfold to more than 1.4 million.

Just as interesting are the monthly statistics, because they show significant variation. For instance, in November 1982, Düsseldorf set a new record for the number of monthly calls; connection was made 64,500 times in the Düsseldorf center. That was 32 percent greater than the number made in March and July 1982, which was the lowest number at that center. Facing peaks in use has always been a challenge to online systems. Still, at first, if and when the Series/1 network starts rolling, text and data load increases can be kept under control thanks to the Bundespost decision to keep the old two-tier system for household communications: 1.2 kbps for receiving information and only 75 baud for sending to the regional switch.

That two-tier system has not been adopted for videotex implementations in the United States, where a uniform send and receive speed is preferred. Though it has been the British solution from the beginning, an American experiment made by GTE under British license has demonstrated that it is ineffective when personal computers are installed in consumers' homes, and the PC has become a household item. A 1.2-kbps send-receive is the minimum capacity to have, according to American design perspectives.

The result of this bottleneck is that, as Fig. 18-1 indicates, Datex-P will be serving the private computer installations in the business sector. Overlooked, however, is the fact that technological advances with microprocessors and memory devices tend to blur the distinction between business and home use and make artificial limits untenable. Furthermore, a connection to Datex-P requires that the user organizations work with packet-switching protocols. Though packet switching is a solution for the future, it is currently adopted by only a small minority of users: The majority of data communications remain of the start/stop type followed, in terms of frequency of use, by the bisynchronous protocols. Even in this present, experimental stage, half the companies participating in the German experiment cannot attach to Datex-P and must be channeled through the videotex lines. In other words, the connection for circuit switching protocols is made directly by videotex.

Though it can be argued that the lack of universality in packet-switching protocols can be remedied through gateways (packet assembly and disassembly), conversions are expensive and may lead to delays and errors. It would be much more efficient if, given the existence of a start/stop network (videotex), start/stop connections were made directly to this network at higher speeds than those presently supported. That is another reason why 1.2-kbps or, better, 2.4-kbps lines should be a must for send and receive. Nor should we forget that, in reality, there is a third technical group of message switching as well: telex. Several years ago, the Bundespost was the first to upgrade the telex lines from the classical 75 baud first to 1.2 kbps and then to 2.4 kbps. And there are plans to hookup the message-switching network with the two others.

Finally, in connection with Fig. 18-2 we observed that international videotex connections will be made in two ways: Those compatible with the packet-switching protocol will be linked to the nearest Datex-P center, whereas international videotex connections following any other protocol, for instance BSC, will be channeled directly to the Ulm center.

Another challenge to the German Post Office is the modems. Contrary to the policy adopted in the United States (after the Carterphone decision) and in England, the continental European telephone companies (PTTs) consider the modem to be an integral part of their monopoly. The Bundespost says that it has to be kept that way because the modem integrates the user's identification number.

That is only a minor argument. The major issue is suppression of possible technical developments. Technology demands that a clear-eyed policy be followed with modems. Any monopoly deprives the user of the fallout of technological developments, and such developments now come quite fast. (There are, however, indications that the Bundespost policy may change by 1986 or 1987.)

Range of Services and the End User

Not only does the adopted standard present significant differences between German and American banks, but that is also true of the range of services: Whereas German financial institutions particularly promote banking functions through videotex, their American counterparts are more prone to integrate merchandising, news, and travel and there is considerable talk about the wisdom of extending the coverage to insurance. The point here is that there are enormous implementation opportunities with videotex, and insurance is as good an opportunity as any. As the services that are offered multiply, we can envision the time when everybody has a PC on his desk (in his office) and in his livingroom (at home). Through it he can automatically access rich and updated databases.

It is feasible to offer the television set, with or without microprocessor support, as an access device to the consumer. For a small fee, when we purchase a new color television set, the modem for the videotext service can be built into it. A cost-effective means of access will promote a new service industry. Information providers will develop complete packages including responsibility for signing up the service, running the total system, and doing marketing.

There is another way to look at this issue: As an easily accessible, efficient, interactive medium, videotex makes sense. Implementation over a wide range will take place in the next 5 years, and it is not particularly important to tackle all product offerings at the same time. What is important is to get started in order to gain experience. It is just as vital to always remember that the videotex service is offered for the end user. That should be written in block letters both by the system managers and by the information providers.

The German information provider controls the content of the pages he places on the system. The end user (Endverbraucher) accesses that information and/or contributes his own. Statistical projections based on the current applications experiments indicate that some 90 percent of the queries posed to the VDB are local. Each participant must have an individual access identification number. This is transmitted automatically to the videotex center during the call setup phase. In addition, the participant may agree upon a four-digit personal password which precludes unauthorized use by third parties.

Because of the facility of database access, the IP should always bear in mind that the videotex user can retrieve any information being offered. Therefore, text, data, and graphics should be adopted to the requests and interests of the user. The risk of negative advertising as a result of inadequate presentation is far greater than with the more classical media.

In regard to the population, the goal of the Bundespost is to have, by 1986, 1 million subscribers: companies as well as private. To achieve that goal, an imaginative tariff policy is being followed:

- The monthly rental for the modem is DM 5 (about $2).

- The user pays 23 pfennigs (10 cents) every 8 min in West Germany for online connection and 23 pfennigs with no time limit in West Berlin.

This tariff has been instrumental in increasing the utilization rate of Bildschirmtext. However, the lack of a time limit in Berlin also has adverse effects on port availability: There are roughly 2000 end users and IP per 100 ports supported at the regional center. If 100 subscribers leave their machines on "because it costs nothing" they saturate the ports.

Let us emphasize the lesson: Tariffs must be studied very carefully to promote the videotex service, offer an income to the telephone company and the IP, and at the same time discourage abuse. Abuse can come from many quarters. For example, the telephone company may overcharge for its service and thus work against its own best interests.

Certainly the organization promoting videotex makes an investment and should, therefore, also make money. But too high a price, like that charged by the Italian SIP (the telephone monopoly), is sure to kill the product in its infancy: What consumer would pay 45,000 to 75,000 lire per month (roughly $30 to $50) for a service the telephone company tries to push down his throat?

The Cost of Standards

In sum, the experience has been positive and the banks participating in it implemented, or intended to implement within this evolving concept, four basic online services for their clientele through videotex:

1. Account balance and associated details

2. Information on deposits, withdrawals, and transfers in general

3. Information on stocks and bonds for portfolio management

4. Payment through Eurochecks and confirmations

The consumer reaction also was enthusiastic. In Berlin there was a waiting list of twice as many videotex applicants as the experiment could accommodate. Suddenly, something changed, and the impact of that change has been so great that it risks upsetting the orderly progression toward videotex use by the German public as a whole.

That something has been the establishment of the CEPT standard. One of the ironic things about standards is that, although they are wanted by practically every user, few calculate the cost of their implementation. Videotex is not the only example. In a conference held in March 1983 with 3M/Interactive Systems, the LAN manufacturer asserted that the implementation of the IEEE 802 Standard will nearly double the cost of the bus interface unit. The reason is that the need to support functions required by 802 (such as priority handling) makes the design of new modems mandatory. Current modems mass-produced for CATV do not support such functions, but they have the advantage of being offered at low cost.

Now it is revealed that, with the CEPT standard, the TV sets and decoders currently used are no good and it will be necessary for manufacturers to develop and users to buy new devices. The hitch is that a TV set conforming to CEPT will cost DM 2000 and a decoder DM 1000. Who will be willing to pay the DM 3000 extra cost?

Although Japanese manufacturers in Germany talk of being actively at work on a black box that will be plugged between modem and decoder and thus save the old hardware at the cost of DM 350 ($140), such a device is not yet available. If the cost stays at DM 3000, what might be the effect on the projected population of users, which, by German estimates, may include 600,000 companies and 400,000 private people by the end of 1984 and early 1985. Will the extra cost mean a death sentence for videotex?

The Bank für Gemeinwirtschaft conducted some market research to answer that question, and the results are the following:

1. The prospective cost will not inhibit companies from adopting videotex usage.

2. The prospective cost will subsequently reduce the number of private users with one notable exception.

3. The exception centers on the professionals: lawyers, physicians, architects, and tax advisers.

The market research by the BfG has established that professionals will pay the extra DM 3000 for rational reasons: To keep the full accounting procedures at the bank. That, plus the fact that the general public will not accept the additional expense, will tend to split the banking population into two broad classes.

German banks are now looking for a strategy to cope with the evolving situation. For example, the Verbraucherbank, which has offered computer-based services to its clientele since 1978 and was the first to go on videotex, is financing its clients (through special credit lines) in the purchase of keyboard and decoder. Many other German banks, among them BfG, have under study a two-tier pricing system for the management of client accounts that will reflect the fact that conventional account management costs too much money. There will be lower charges for accounts working through videotex and higher charges for accounts using the classical method.

Another strategic consideration by the German banks is the merger of home banking, ATM, and point of sales into one integrated marketing philosophy. There also seems to be considerable German bank interest in studying the implementation of the chip in card within the perspective of an integrated marketing philosophy, particularly for identification purposes.

That fits well with the existence in Bonn of a central credit organization (Zentraler Kredit Ausschuss) which works, among other things, on rules for handling online credit including the norms for Bildschirmtext. One of the procedures being elaborated is that the loss of a personal identification number or chip in card will immediately result in all checks for that particular identification being blocked throughout the German banking system.

Conclusion

The information the reader finds in this book was gained through research and experience. For that reason, I kept closely to the fundamentals of organization, user interactivity, and databasing as they apply to whatever type of network may be developed. A basic message to retain is that the old distinctions between computers, information handling, and telephones are now technical nonsense. We are moving toward a new standard. The reporting structure most companies have developed during the last 20 years is typically batch- and hardcopy-oriented. Voice mail and videotex present new processes: The latter is interactive and rests on visual facilities with graphics and color, and both benefit from store and forward.

The key driving force behind the spread of videotex in the home will be the extent to which advertisers are prepared to use it. Broad use will sharply reduce its cost to subscribers, and cost is a key ingredient to the success of any product. Once videotex becomes firmly established, it will have profound effects on seemingly unconnected social areas such as industrial relations, architecture, manufacturing design, and sales perspectives. The shift away from conventional workplaces will change industrial relations and corporate structures. The home will become a place of work, most of which will be done at the videotex computer terminal.

Distinctions between residential and business areas in cities may eventually blur, with a corresponding drop in the demand for office space. Home-based shopping will profoundly alter not only retailing patterns but also manufacturing processes because consumers will be able to detail, through their videotex terminals, the precise specifications of custom-made products.

Technology knows no frontiers, and it can be a friend or foe depending on our attitude. To be competitive, we must feed our opportunities and starve our problems. We have the means to project the coming developments, to assemble diverse technologies into working systems, and manage the integration of networks, of software, of hardware, into a coherent service. *But are we taking full advantage of this capability?*

List of Initial-Letter Abbreviations

ABR Audio-Balance Reporting

ACH automatic clearing house

AI artificial intelligence

ANSI American National Standards Institute

APDU application protocol data unit

ATM automated teller machine

BIS business information system

BO branch office

BPO British Post Office

bps bits per second

Bps bytes per second

BPW bits per word

BSC Bindry Synchronous Communications (bisync) protocol

Btx Bildschirmtext

CAD computer-aided design

CATV cable television

CCD charge-coupled device

CCITT Consultative Committee for International Telephone and Telegraph

CEPT European Committee for Post and Telecommunications

CIF central information file

COAM company-operated and maintained network

COM computer output to microform (microfilm)

CPU central processing unit

cr carriage return

CSA Canadian Standards Association

CUG closed user group

DB database

DBA database administrator

DBMS database management system

DC data communication

DDA demand deposit account

DDB distributed databasing

DDC distributed data communication

DDP distributed data processing

DIS distributed information systems

DOR digital optical recording

DOS disk operating system

DP data processing

DRCS dynamically redefined character sets

DSS decision support systems

DTC depository transfer checking

EEC European Economic Community

EMS electronic message services

EUF end user functions

FBS First Bank System

FE front end

FEP front end processor

FIT foreign exchange transactions

FTD foreign trade data

h hour(s)

HB home banking

HDLC High-level data link control

ID identification

IE information elements

in inch or inches

I/O input/output

IP information provider

IRV International Reference Version

IS index sequential (file organization)

K 1024 bytes, the common unit of storage devices

kB kilobytes

kbps kilobits per second

LAN local area network

MB megabytes

MBps megabytes per second

Mbps million bits per second

min minute(s)

MIPS millions of instructions per second

MIS management information system

MIS/DP management information system and data processing

mm millimeter(s)

MPPS Market Price Display Services

NA/PLPS North American Presentation Level Protocol Syntax

NASA National Aeronautics and Space Administration

NKC numeric key choosing

NTT Nippon Telegraph and Telephone

OA office automation

OS operating systems

OSI open system interconnection

PASS portable analysis and synthesis speech

PBX private branch exchange

PC personal computer

PDI picture description instruction

PIC personal identification code

PIN personal identification number

P/L profit and loss

PLP presentation level protocol

PLPS presentation level protocol syntax

PMI picture manipulation instructions

POS point of sale

PPCI presentation protocol control information

PPDU presentation protocol data unit

PSDU presentation service data unit

PTT postal, telephone, and telegraph service

RET real enough time

RPG report program generator

RT realtime

s second(s)

SI shift in

SO shift out

TBP telephone bill payment

TDB text and database

TKC two-key choosing

TS time sharing

UID user identification

URF user response facility

VAN value-added networks

VDB viewdatabase

VLP video long play

VSF voice store and forward

WP word processing

WS workstation

Glossary

Access techniques Techniques that permit nodes to gain control of the channel to transmit messages; they must be implemented in relation to a network topology.

Active position The position on the screen from which subsequent actions would take place if they were activated.

Analog signals Signals that vary in a continuous manner and can be seen as having an infinite number of states.

ASCII (American Standard Code for Information Interchange) A 7- or 8-bit code that provides up to 128 or 256 different characters, respectively. Originally, an eighth bit could be added as a parity check for error detection. With PLP, the eighth bit can be part of the code (G-sets, C-sets).

Asynchronous transmission The transmission mode by which characters may be sent with random timing. The data bits of each character are introduced by a start bit and followed by a stop bit. The asynchronous mode is common for low-speed transmission, less than 2.4 kbps.

Attributes display A means of modifying the presentation of characters on the screen. Attributes may be applied to the full screen, a full row, part of a row (serial), or to subsequently printed characters (parallel).

Background color The color of the area of the character cell not occupied by the foreground color; the color of the remaining area of the character. The color may be any from the available color tables or be transparent, in which case the full screen background color (or the cumulative result of all picture elements previously set or the video picture) is seen.

Baud A unit of digital signaling rate equal to the reciprocal of the length, in seconds, of the shortest signal element. The information rate in *bits per second* (bps) may be greater than the baud rate, because one signaling element can represent more than one bit. Usually dibits and tribits are created at the modem level through the division of frequencies.

Bit combination An ordered set of 7 or 8 bits (binary digits) that represent a character. (*See* **ASCII.**)

Bit error rate (BER) A measure of the performance of a digital transmission system; it expresses the probability of error per bit transmitted.

Block error rate The probability of one or more errors in a block of bits of specified length. Still another unit of measure is the percent of *error-free seconds*.

Border area That part of the display screen (visible display) which is outside the defined display area.

Bus topology Like a multidrop line, a bus topology is shared by a number of nodes. Bus architectures have most frequently been used for peer systems, with messages being broadcast to all nodes. Each node must be able to recognize its own address in order to receive a message.

Cable plant The system of wires in a building. For data communications purposes, the cable plant will typically be made of coaxial cable, twisted pair, or other wire. Eventually, it will be made of optical fibers.

Call detail recording A private branch exchange feature that creates a log of every telephone call and session so that use patterns can be determined and charges made on the basis of services rendered.

Carrier-sensing multiple access (CSMA) The ability of each node to detect traffic on a channel and act accordingly. This is a contention method.

CCITT (International Telephone and Telegraph Consultative Committee). The members of this international committee are the post and telecommunications authorities of each country. The United States is represented through a special office of the State Department. Though CCITT has consultative character, its "recommendations" in the field of telecommunications stand a good chance of becoming the international standard. A rather recent example is the X.25 recommendation for packet switching.

Character A member of a set of elements that is used for organization, control, or representation of data. A character repertoire contains two types of elements: graphic characters and control functions. Graphic characters can be alphanumeric, block mosaic, smoothed mosaic, or line drawing. Accented characters are handled by using the composition method of coding. The fixed repertoire of characters can be extended with dynamically redefinable characters.

Character display Data sent to the terminal and used to generate alphamosaic displays in which text and graphic characters are displayed, usually in a fixed format of rows and columns.

Circuit switching Switching that involves the establishment of a total path at all initiations. The path is established by special signaling which threads its way through different switching centers and is subject to the speed and code limitation of the weakest link.

Closed information Videotex information which is intended for only a restricted user audience and to which access can be had only by the use of a key (code) number. It may consist of company confidential information or text, data, and images of high current value sold to subscribers. The concept makes a closed user group (CUG) possible.

Code table A table showing the character corresponding to each bit combination in a code. Normally it is presented as a rectangular matrix of columns and rows.

Coded character set A set of unambiguous rules that establish a character set and the one-to-one relations between the characters of the set and their bit combinations.

Compunications The blending of voice and data communications with computers. It can be put to three main types of use: replacing lengthy telephone calls, substituting for short memos, and introducing new ways of communication which were not possible before.

Conceal To display characters as spaces until the user chooses to make them appear.

Control character A control function the coded representation of which consists of a single bit combination.

Control function An action that affects the recording, processing, transmission, or interpretation of data. The coded representation of a control function consists of one or more bit combinations.

Data communication End-to-end transmission of information other than voice, music, or video.

Data source Information in digital or analog form. Data transmission handles any data source; digital transmission is a subset characterized by digital implementation.

Data syntax The format, rules, and procedures for the encoding of alphanumeric text and pictorial information contained in a presentation service data unit (PSDU).

Data transparency A means of representing all possible 7-bit or 8-bit combinations, respectively 128 or 256, as data without risk of confusion with management commands or the header.

Decision Support Systems (DSS) Systems that identify computer-based algorithmic approaches, transparent to the user, which operate on text and databases to present to senior management (preferably in a graphical form) the information needed for important decisions. The trend in DSS is to answer questions and work on problems as yet undefined—precisely within the unstructured information environment characterizing the executive level.

Defined display area The rectangular position of the display in which all text and pictorial images may be presented.

Device control functions Functions that include the action of scrolling, the display of the cursor, others which may be controlled by codes transmitted to the terminal.

Digital contention switch A computer accessory that moves data parallel with a voice network. It must be supported by a lot of planning and coordination.

Digital signals Coded information having a limited number of discrete states prior to transmission.

Direct inward dialing A means by which an outsider to a private branch exchange can dial a number and ring the telephone on a specific desk without the assistance of a telephone operator with its attendant cost and delay.

Directory A dynamic database maintained in a modern private branch exchange that allows the user to determine, during call setup, an account code, the dialing privileges the user enjoys, and the priority to be associated with any calls that must be queued. It further permits maintenance and diagnostic programs to know the exact configuration of devices on each desk and the cable path from the switch to those devices.

Editing terminal A terminal designed for use in projecting, preparing, or modifying videotex pages. It is distinguished by its facilities for encoding alphanumeric, color, and graphics information. Online editing terminals operate directly to the viewdatabase computer. Offline (or local) editing terminals permit the preparation, viewing, and modification of videotex pages (both routing and information) on a local basis.

Electronic mail (Email, Teletex) A message system that operates at normal voice grade line speeds, rather than the slower telex circuit line speeds, with a full alphanumeric character set. (Teletex is the CCITT term for electronic mail).

Erlang A unit of traffic intensity that has no dimension and is employed to express the average number of connections underway or the average number of devices used on a line. It reflects the capacity of a communications system. Traffic in erlang is the sum of the holding time of paths divided by the period of measurement. One erlang expresses the continuous occupancy of one traffic path.

Erlang B A traffic model used by AT&T. It is based on the assumption of Poisson input, negative exponential holding, and blocked calls cleared.

Erlang C A queuing model used by AT&T. It follows the erlang B assumptions but with blocked calls delayed. The servicing discipline is approximately first come-first served.

Final character In videotex, the last character of a management command of presentation protocol control information.

Flash In videotex, to vary in color at regular intervals. In normal flash *the characters are displayed alternately in the prevailing foreground color and in the prevailing background color. In inverted flash* the colors are changed on the inverted phase of the flashing clock. In *reduced intensity flash* the characters are displayed alternately in the prevailing foreground color and the equivalent color of the next color table: table A colors adopt table B colors.

Foreground color The color of the graphics shape that is being displayed in a character cell. It may be any color from the available color tables or be transparent. In the latter case the full screen background color, the cumulative result of all picture elements previously set, or the video picture is seen.

Format effectors Control functions that influence the positioning of text and pictorial images within the defined display area on a presentation device. Characters may be positioned within the defined display area by means of format effector controls which move the active position, usually in units of one character position.

Frame The smallest unit of retrievable information. It is a segment of a signal (analog or digital) with repetitive characteristics, so that corresponding elements of successive frames represent the same thing. Each frame generally contains bits specified for source and destination addresses, control, data, and parity.

Framing The process of establishing a reference so that the elements of a frame can be identified.

Frequency division multiplexing (FDM) A transmission technology employing different frequency bands with a means, at each end, of discriminating and recognizing the assigned frequencies.

Graphic character A character, other than a control function, that has a visual representation normally printed or displayed.

Graphic code extension The method of encoding more graphic charactrs than can be represented by the code combinations of the basic code table. Alternative sets of characters can be designated by means of shift functions.

Home position The position on the screen from which subsequent actions would take place if they were activated.

Home videotex The use of videotex in the home environment. It can be helped by home computers. Early examples include such applications as TV games and chess.

Hub A device at which a branch of a multipoint network is connected. A network may have a number of topologically distributed hubs. Also called bridging point.

Information provider (IP) An organization that provides information for storage on a viewdatabase. Typically, IPs are companies that rent space on the public utility system which can be called up by videotex subscribers and for which a charge is made by the organization running the videotex service. After deduction of operating costs, the resulting revenue is passed on to the IPs concerned.

In-house system A videotex system operated by a single information provider on which to display his own text, data, and images. (Nothing prevents the IP from also renting space on his in-house system.) An in-house system can be supported in one location or, in long haul, by leased or public telephone lines.

Intermediate character (I) A character used preceding a final character to increase the number of possible management commands

Invert operation A display of characters such that foreground and background colors appear to have been exchanged. If flash is applied, the polarity of the flashing clock also is inverted.

Lining A display of alphanumeric characters with an underline that is considered to be part of the shape of the characters. Mosaic characters and line drawing characters are displayed in separated fonts.

Management command A parameter value P followed by a command identifier C that represent a presentation level management action such as a change from one data syntax to another.

Markers Flags in a memory to show where attribute controls have been set. They are associated with the leading edge of the character position.

Marking Identifying characters for further action at the terminal, as to be transferred to an output device. More than one type of marking is possible, and each may be separately processed.

Message switching One of three communications disciplines: circuit, message, and packet switching. Telex (TWX) and Swift are examples. Because of store and forward capabilities, it requires no direct wire connection. It permits broadcast of messages; delayed delivery is feasible if recipient is not available or lines are busy; optimization is in terms of throughput; but the network and its functions are applications-oriented.

Modem A device to convert a digital signal to an analog signal that can be transmitted across a voice grade circuit to another modem that reverses the process. This modulator-demodulator modulates the telephone transmission frequencies at the sending side of the message and demodulates them at the receiving end. Modems make it feasible for information to be moved from one place to another at very low energy cost. The switching of frequency channels within a band is accomplished by *frequency agile modems.*

Multipoint link A single line shared by more than two nodes. Also called a multidrop link.

Network architecture The relations and interactions between attached devices.

Node A basic information-handling unit which is directly addressable and is attached to a network.

Nonblocking An architectural property of computer-based switches that allows 100 percent of the installed devices to be active simultaneously and minimizes the extent of traffic engineering required.

Numeric parameter A parameter the value of which is a decimal number.

Office automation The process by which a wide variety of productivity-related products is into a computer-based system for office work. It ranges from simply bringing together copying, text input, and text handling and distribution to providing the necessary understructure for data, text, image, and voice.

Open information Text, data, and images accessible to all videotex subscribers; hence, public information is open information. Examples are train and airline timetables, restaurant guides, and general news.

Open System Interconnection (OSI): A system model for organizing interchange protocols into seven functional layers of abstraction and representation.

Packet switching Transmission and switching of packets obeying a well-defined protocol. Messages can be split into packets or, alternatively, regrouped into packets depending on their length. Like message switching, there is no direct wire connection in packet switching; but unlike message switching, the network and its facilities are applications-independent. Packet switching permits broadcast of messages; it can employ narrow or wide bandwidth according to need; speed and code conversion are feasible; the route is dynamically established for each transmitted packet; and reliability is reasonably high.

Parallel attributes The property of the active position that move with it under the action of format effectors or spacing display characters (including space). They apply to the displayed characters subsequently received until they are changed by relevant controls including certain format effectors.

Parallel transmission mode That which sees to it that the bits of a single message are sent at the same time; the whole message is transmitted in the time it takes to send one bit.

Parameter value (P) One or more characters with videotex that represent a numeric or selective parameter whose interpretation depends on the particular command with which it is associated.

Point-to-point connections Standard I/O channels that connect a central resource to each of its terminals and vice versa. They limit distance and total connectivity while increasing cost and complexity.

Polling A noncontention method of network access that determines the order in which nodes can take turns accessing the network.

Presentation entity The active element of a terminal that is concerned with presentation. It interacts directly only with elements in the next higher division and the next lower division within a host or terminal.

Presentation process The activity within one presentation entity that interprets the semantics of a particular presentation data syntax.

Presentation protocol control information (PPCI) In videotex, a header for a presentation protocol data unit (PPDU) that serves to delimit the PPDU from preceding and following PPDUs.

Presentation protocol data unit (PPDU) A unit of data consisting of presentation protocol control information (PPCI) followed by a presentation service data unit (PSDU).

Presentation service data unit (PSDU) The portion of a presentation protocol data unit (PPDU) that contains either management commands or a particular data syntax as indicated in its PPCI.

Prestel A viewdata service operated by the British Post Office (BPO).

Protecting The process of assuring that character positions are protected against alteration, manipulation, or erasure. The protection is valid for attributes as well as characters. Protected characters must not be obscured by enlarged characters. Protected character positions may be overwritten only by the use of a specific code or by the action of the clear screen command (CS), which deletes both the characters and the protection. Such protected character positions may be scrolled and therefore may disappear from the screen, because the protection is always related to the particular information on the screen.

Public utility systems Among others, videotex systems operated by organizations whose primary role is to provide a service able to support a variety of information providers in displaying their information and a large number of users. Practically, any telephone user can become a subscriber.

Ring topology A process whereby the connected nodes form an unbroken circular configuration. A typical ring is a peer system. In a manner similar to that in a bus structure, each node in a ring topology must be able to recognize its own address. It can also serve as an active repeater retransmitting information addressed to other nodes. Rings with centralized control are referred to as *loops*. Loops may also be designed in such a way that all message exchanges pass through the control node.

Scrolling area The portion of the defined display area within which the characters and associated attributes move in increments of one character position under the action of format effectors or specific controls.

Selective parameter A parameter the value of which represents an item in a list associated with the particular management command.

Serial attributes Attributes set between markers on a row. They apply from the position active at the time they are received to the end of the row or until a contradictory marker is reached.

Size One of the basic parameters in videotex. There are four states of character size: *Normal-size* characters occupy the active position. *Double-height* characters occupy both the active position and the corresponding position of the adjacent row. *Double-width* characters occupy both the active position and the next position of the same row. *Double-size* characters occupy the active position, the next position on the same row, and the corresponding two positions on the adjacent row.

Space division multiplexing (SDM) The oldest of the transmission technologies; it is associated with the physical metallic path which exists in a given switching connection.

Synchronous transmission Transmission characterized by a constant time interval between characters determined by a special digital clock that is usually located in the modem. Start and stop bits are not needed, and data transmission is therefore more efficient. Synchronous transmission is generally at speeds of 2.4 kbps or more.

T-carrier The AT&T long-lines digital carrier system that will transmit at 1.544 Mbps. It is usually divided into 24 voice channels of 64 kbps, each of which is used to carry voice in an eight-level code sampled at 8000 samples per second.

Telematics A synthesis of functions: the integration of voice communications with text, image, and data communications into a new system. In a broader sense it also includes entertainment and the postal service.

Teletext A *one-way* system that is basically a broadcast message service. It uses the TV signal as carrier and is based on a standard character set which can be received and interpreted by a normal TV set adapted to that purpose. (The CCITT term is broadcasting videotex.) Text and graphics information is transmitted with the television signal. Some 250 pages can be transmitted in sequential and repetitive form at rate of some four pages per second. At the receiving end, the domestic user can select the required page by means of a simple numeric keyboard, such as an adapted TV remote control unit.

Terminal In videotex, the presentation entity that exchanges coded bit combinations with the host by means of telecommunications and presents it for human consumption.

Theoretical terminal model An ideal, perfect terminal. The videotex service, in its alphamosaic option, can be so described.

Time division multiplexing (TDM) A transmission technology employing time segments for channel separation. It is used mainly with wideband carriers when a large number of subscribers are to be accommodated. With TDM, a frame is a sequence of time slots repeated at the sampling rate. Each channel occupies the same sequence position in successive frames.

Token A bit pattern circulating around the carrier, whether physical or logical ring.

Transceiver A transmitter and receiver.

Transparent mode a mode in which data transparency is achieved by using a byte-stuffing mechanism.

Unconstrained topology A nonspecific topology that takes the shape of the actual connections and varies from one implementation to another. The network architecture defines the relations and interactions between attached devices. Supported functions obey protocols and are served by common interfaces. Unconstrained configurations can be made of combinations of point-to-point and multipoint links. The connections are usually determined by economics. Also called a hybrid or mesh.

Videotex An interactive message service (two way, send/receive) using a standard character set which can be controlled and received over voice grade telephone line, and interpreted by a normal TV set or other viewing device with adapted circuitry. The adapted television set includes the line modem and an automatic dialer to establish contact with the videotex computer directly. Once contact has been made, an introductory page is sent back to the receiver.

Videotex charges Charges for the system; they include the telephone line and a cost per page based on the value of the information. Charges are higher for specialized data; there is no charge for menus, classified advertisements, and some of the infopages.

Videotex control functions repertoire The total range of controls which may be communicated between videotex services.

Videotex default The default format: 24 rows of 40 columns with automatic wraparound on rows and columns.

Videotex features The ability to restrict access via security keys to parts of the database and only from authorized terminals; automatic billing of the charges to the user; credits to the information providers concerned; the possibility of making purchases via the system; and message handling facility.

Videotex page A single frame (screenful) of information, containing 24 lines each of 40 characters of the standard videotex character set. Normally only the middle 20 lines are free for information. The first and last line are reserved for the videotex system, and it is well to leave a buffer line on each side.

Videotex predefined graphic character repertoire The total range of predefined graphic and mosaic characters which may be communicated between videotex services.

Word A set of digital signs expressing information in an orderly, preestablished manner.

Wraparound controls In videotex, a set of rules which govern what happens when the active position attempts to move off the defined display area.

About the Author

Dimitris N. Chorafas has been a corporate consultant to the leading banking and manufacturing industries in 50 countries for the last 20 years. A Fulbright Scholar, Dr. Chorafas received his doctorate in mathematics at the Sorbonne in Paris. He is also a graduate electrical and mechanical engineer and has a Master of Science degree in computers from the University of California, Los Angeles. He has written numerous technical articles as well as 35 books published in 14 languages.